Life Cycle Costing for Construction

Life Cycle Costing for Construction

Edited by

JOHN W. BULL
Department of Civil Engineering
University of Newcastle upon Tyne
Newcastle upon Tyne

Taylor & Francis
Taylor & Francis Group

LONDON AND NEW YORK

First published by
Taylor & Francis & Professional in 1993
2 Park Square, Milton Park, Abingdon, Oxon OX14 4RN
52 Vanderbilt Avenue, New York, NY 10017, USA

First issued in paperback 2020

*Taylor & Francis is an imprint of the Taylor & Francis Group,
an informa business*

© Taylor & Francis, 1993

Typeset in 10/12pt Times New Roman by Thomson Press, New Delhi, India.

A catalogue record for this book is available from the British Library

ISBN 13: 978-0-367-57993-7 (pbk)
ISBN 13: 978-0-7514-0056-4 (hbk)

Preface

The construction industry is becoming increasingly aware of the need to adopt a holistic approach to the purchase, design, building and disposal of structures. With some 60% or so of the total construction budget in most developed countries being spent on repair and maintenance, there is a critical need to design for durability and reliability, with carefully planned finance, maintenance and repair scheduling. An important facet is to consider how all costs are allocated and distributed during the lifetime of a structure. This approach, known as life cycle costing, which has the aim of minimizing total lifetime expenditure, is the focus of this book.

In the construction industry the capital cost of a structure is almost always kept separate from the cost of maintenance and from the cost of disposal. It is common practice to accept the cheapest capital construction cost and then to hand over the structure for others to maintain. This may have been acceptable practice when the expected service life of a structure was 80 to 200 years, or where the structure (such as a cathedral) was likely to remain substantially unaltered during its lifetime. It is no longer acceptable practice.

This book looks at a variety of building structures and discusses how, in the initial design of the construction, the cost areas may be considered such that the 'life cycle cost' of the structure may be reduced to ensure a structurally efficient building which will satisfy financial, business, personal and environmental requirements. The book includes chapters on the introduction to life cycle costing, how life cycle costing can be used as a decision tool and how it could have improved existing costing. Further chapters relate life cycle costing to reliability-based and optimum design, the refurbishment of buildings, and highway costing. Discussions of how life cycle costing is used in the defence industry and in the health service are included to show the change in thinking that will be required in the construction industry.

Finally, I would like to thank the chapter authors for their efforts. I also thank my wife Sonia for her help and support.

JWB

Contents

4 Life cycle costing of highways   53
R. ROBINSON

5 Life cycle costing in the defence industry   86
M.J. KINCH

6 The quality approach to design and life cycle costing in the health services

J.F. McGEORGE

7 How life cycle costing can improve existing costing

A. ASHWORTH

Contributors

Dr A. Ashworth 186 Moor Lane, Woodthorpe, York YO2 2YZ, UK

Dr J.W. Bull Department of Civil Engineering, University of Newcastle upon Tyne, Newcastle upon Tyne NE1 7RU, UK

Mr S.J. Dale Crown House Engineering, Crown House, 550 Mauldeth Road West, Chorlton-cum-Hardy, Manchester M21 2RX, UK

Mr J.J. Griffin Systems Analysis Europe Ltd, Scotsgrove House, Gong Hill Drive, Farnham, Surrey GU10 34G, UK

Dr S.-I.K. Gustafsson IKP/Energy Systems, Institute of Technology S-581, 83 Linköping, Sweden

Dr M.J. Kinch 32A Chiltern Road, Hitchin, Hertfordshire, SG4 9PJ, UK

Dr K. Koyama Department of Civil Engineering, Shin Shu University, Wagasato, Nagano 380, Japan

Mr J.F. McGeorge Department of Civil Engineering, University of Cape Town, Randebosch 7700, South Africa

Dr R. Robinson Rendel, Palmer and Tritton, 61 Southwark Street, London SE1 1SA. UK

1 Introduction to life cycle costing

S.J. DALE

1.1 Introduction

Nowadays we all try to forecast the consequences of our decisions before we proceed although we also know that the odds against a forecast being correct are quite formidable. As the overwhelming majority of people in the construction industry are dealing with someone else's money it is usual to utilise some form of accepted forecasting method to predict a result.

While the forecast may not be accurate, if suitable parameters are used and all the interested parties concur with the forecasting method, then it is possible that the ultimate psychological goal, that of confidence, which is a prerequisite to investment, may be achieved. If then, by some mischance, the end result is not that intended, at least it can be shown that the decision was based on knowledge at the time.

Most disagreements over finance are a result of a person's desire not to part with their money. The aim therefore should always be to buy low and sell high. However, more can be paid in the beginning knowing that in the ultimate result the deal will prove to be cheaper or will reap a greater reward. In this event some form of financial analysis of a particular decision becomes necessary. The idea that the economic consequences of a decision can be analysed is logical, and the fact that such an analysis should be recorded for future reference is obviously sensible.

Life cycle costing is a mathematical method used to form or support a decision and is usually employed when deliberating on a selection of options. It is an auditable financial ranking system for mutually exclusive alternatives which can be used to promote the desirable and eliminate the undesirable in a financial environment.

As decision-taking lies at the heart of all our working hours, aids to the taking of decisions are valued and are used to justify our actions. An ability to forecast the consequences of our decisions eliminates uncertainty and forms the basis for ultimate success.

For example, structural engineers know, from the results of laboratory experiments, that a certain size of steel member will support a given load. They can point to the laboratory record as a reason to justify their actions. Usually, the laboratory record will provide several solutions, or sizes of member to support the load. Unsuitable solutions can be eliminated on the basis of spacial requirement, availability, constructional difficulties or cost. In forming the decision on which steel member to select, the engineer needs to record and justify the final selection. This enables the decision to be audited for effectiveness at some future date.

Usually, the subject of 'which solution is the cheapest' has to form part of this decision-making process, and a method of supporting a financial decision needs to be established and recorded to justify such a decision. Life cycle costing can be described as a means of auditing the financial consequences of a decision.

1.2   The problem

In Table 1.1, five solutions to the specification 'build an office block' have been designed and costed. Solution B has obvious financial benefits and therefore will be the selected option. This 'lowest-cost' method of decision-making is, without question, the current major method of building option selection and works on the assumption that the cheapest solution is the best financial option.

During the 1930s many building users began to discover that the running costs of the building (i.e. maintenance, energy, management, etc.) began to impact significantly on the occupiers' budget. It was found that the 'lowest-cost' system of selection was not always the cheapest solution over the lifetime of the building. It became obvious that some other method of financial analysis which takes into account the running (or resource) costs of the building must be used to give credence to the decisions when a number of options are under consideration.

Table 1.2 expands the data and changes the decision on the building selection. Over the life of the building option A appears to be the cheapest solution. However, the basis for this decision does not stand up to close inspection. We all know that if maintenance costs are £400 000 in the first year of a building's life they are unlikely to be £400 000 in the tenth year of life. This is due to a number of factors such as inflation, replacements, etc. Other factors may also come into play, such as a shortage or glut of raw

Table 1.1

Building	Construction cost (£)
A	10 M
B	8 M
C	15 M
D	9.5 M
E	11 M

$M = 10^3$

Table 1.2

Building	Capital cost (£)	Maintenance costs/annum (£)	Life-span (years)	Demolition costs (£)	Simple lifetime costs (£)
A	10 M	400 000	30	100 000	22.1 M
B	8 M	500 000	30	100 000	23.1 M
C	15 M	300 000	30	100 000	24.1 M
D	9.5 M	500 000	30	100 000	24.6 M
E	11 M	425 000	30	100 000	23.85 M

$M = 10^6$

materials, which would change the base price of a product and could modify the replacement specification (e.g. from timber to aluminium). Similarly, items may require periodic change, over a number of years, resulting in a variable annual maintenance charge. It is clear that a system of financial evaluation that can make allowance for all variables throughout the life of the building and reduce the options to a simple single-figure selection, as in Table 1.1, would make decision-taking a great deal easier.

1.3 The methods

Several options are available. All are well documented and have been in use, certainly since the early 1930s, in many different business sectors. The three most commonly used in the building sector are as follows.

1. *Simple payback*: defined as the time taken for the return on an investment to repay the investment.
2. *Nett present value*: defined as the sum of money that needs to be invested today to meet all future financial requirements as they arise throughout the life of the investment.
3. *Internal rate of return*: defined as the percentage earned on the amount of capital invested in each year of the life of the project after allowing for the repayment of the sum originally invested.

All three methods are accounting systems developed initially for the manufacturing industry to determine the financial worth of an investment. The three methods have been developed to determine if an original investment is worthwhile. For example, the purchase price of (or investment into) a new machine may be £1 million, but extra income earned by the resultant cheaper manufacture, increased production or higher product quality from the machine may be £200 000 per annum.

Here the investment generates a known return. In building we generally wish to know if additional money spent on the construction of a building is worth the savings that will be made by a subsequent reduction in running costs. For example, the specification of ceramic tiles in a toilet may be more expensive but the saving in maintenance costs over the alternative painted surface may prove worthwhile.

1.3.1 Simple payback

Simple payback is a simple method of cost appraisal used by many in industry, particularly to evaluate energy-saving schemes. Simple payback can be expressed as:

$$P = I/R \qquad (1.1)$$

where P = payback period (years), I = capital sum invested, and R = money returned or saved as a result of the investment.

Thus for our machine example

$$\text{Payback period} = \frac{1\,000\,000}{200\,000}$$

$$= 5 \text{ years}$$

The decision process now has to decide if five years is an acceptable period for a return on investment capital.

Applying this principle to the options in Table 1.2, it is necessary to adopt a comparator approach to compare solutions A and D. Is the additional investment of £500 000 for solution A worth the additional annual saving of £100 000 in maintenance costs?

$$\text{Payback period} = \frac{(10\,M - 9.5\,M)}{(500\,000 - 400\,000)}$$

$$= 500\,000/100\,000$$

$$= 5 \text{ years}$$

In making a decision to purchase option A rather than option D, it is necessary to assess whether a five-year return on the additional investment is worthwhile.

The use of simple payback is limited by its result. An evaluation of the acceptable payback period is necessary, for which no method or criterion is shown or established. In practice, a maximum period of two or three years is set as a criterion for investment. This is primarily due to the current vogue for a quick return on an investment or 'short-termism' but also because the calculation makes no allowance for the following variables:

- Inflation
- Interest (payable or received)
- Cash flow
- Taxation

Reducing the payback period is thought to limit the likely effect of these factors; however, this may not be the case. Considering the effect of taxation alone on the simple payback method shows a significant variation in the result. Business expenditure (e.g. maintenance charges or resource costs) attracts a 100% allowance on UK corporation tax, which is currently set at 35%. A saving in the cost of maintenance will result in a subsequent loss of relief from corporation tax; therefore the value of the saving is reduced. A resource cost of £100 per annum with relief from corporation tax at 35% will only cost £100 − (£100 × 0.35) = £65 per annum. Therefore a visible saving of £100 per annum is in fact only worth £65 per annum.

The capital expenditure, or original sum invested, is also affected by taxation. For example, assuming that the capital sum invested attracts relief under the range of capital and revenue allowances (for plant and equipment, say) of 25%, on a written-down balance over four years tax relief is:

Year 1	$500\,000 \times 0.25 \times 0.35$	$= 43\,750$
Year 2	$500\,000 \times 0.75 \times 0.25 \times 0.35$	$= 32\,810$
Year 3	$500\,000 \times 0.25 \times 0.25 \times 0.35$	$= 21\,870$
Year 4	$500\,000 \times 0.25 \times 0.25 \times 0.35$	$= 10\,940$
	Total relief	£109 370

Therefore, actual capital cost = £500 000 − £109 000
$$= £391\,000$$
Relating this back to the simple payback formula

$$\text{payback period} = \frac{391}{65} = 6 \text{ years}$$

If the building is located in an Enterprise Zone in the UK, a 100% capital allowance is applicable, which will reduce the payback period to one year.

In preparing this calculation it is assumed that the building owner is in a position to pay sufficient corporation tax to receive the full benefit. If the original investment funds are the subject of a loan, repayable at a fixed or variable interest rate, then the capital cost will rise. Also, the £100 saving may be further reduced because payment is made after one year, during which inflation may have eroded the value of the saving. The result is to increase the payback period.

The use of the simple payback method is therefore fraught with difficulties if used by decision-takers without full knowledge of the financial circumstances of the investor. However, numerous operational decisions have been based solely on this method in order to determine the viability of options.

The simple payback technique does have a role to play. The quick elimination of unrealistic options, which have a result of over 10 years, is useful in that complex calculations can be avoided by using simple payback as a coarse filter of solutions at an early stage. Certainly simple payback should not be used more extensively.

1.3.2   Nett present value

The nett present value (NPV) of a flow of cash is a system proposed by many as the best for evaluating building-related options. The system takes into account all the apparent variables acting upon a cash stream. Flannagan *et al.* (1989) express nett present value as:

$$\text{NPV} = \sum_{t=0}^{T} \frac{Ct}{(1+r)^t} \tag{1.2}$$

where C is the estimated cost in year t, r is the discount rate, and T is the period of analysis in years.

The discount rate is a method of determining the time value of money. For example, £100 invested today at 11% per annum will be worth £111 in one year's time, or:

$$T = \text{PV}(1+r)$$
$$= 100(1 + 0.11) \tag{1.3}$$
$$= 111$$

where T is the value at one year, PV is the original investment or present value, and r is the interest or discount rate.

If we wish to know how much to invest today to meet a cost at some future year, the formula becomes:

$$\text{PV} = T/(1+r)^n \tag{1.4}$$

where n = number of years.

Considering an example in which a sum of money is to be set aside to pay for a annual expenditure stream of £100 per annum for five years, when invested at 11% interest this becomes:

$$\text{PV} = \text{Year } 0 + \text{Year } 1 + \text{Year } 2 + \text{Year } 3 + \text{Year } 4$$

$$\text{PV} = T_0 \frac{T_1}{(1+r)} + \frac{T_2}{(1+r)^2} + \frac{T_3}{(1+r)^3} + \frac{T_4}{(1+r)^4} \tag{1.5}$$

$$\text{PV} = 100 + \frac{100}{1.11} + \frac{100}{(1.11)(1.11)} + \frac{100}{(1.11)^3} + \frac{100}{(1.11)^4}$$

$$= £410.24$$

The formula allows for the expenditure to vary from year to year, allowing for differing intervals of replacement of equipment.

The above example makes allowance for interest receivable on the sum invested. In reality, the value of our investment will be eroded by the pernicious effects of inflation. The formula, equation (1.5), can be modified to a factor which will take account of inflation. Inflation will increase the costs at year

n and therefore increase the present-day investment level. The modified factor is known as the 'nett of inflation discount rate' the (ndr) where

$$ndr = \left[\left(\frac{(1 + \text{interest\%})}{(1 + \text{inflation\%})} \right) - 1 \right] \qquad (1.6)$$

If inflation is 6% per annum and interest is received at 11%, then

$$ndr = \left[\left(\frac{(1 + 0.11)}{(1 + 0.06)} \right) - 1 \right]$$

$$= 4.7\%$$

Substituting this modified discount factor into equation (1.5), the cash reserve will need to be £457.03.

This figure represents the sum of money needed to be invested today, at an interest rate of 11% with inflation at 6%, in order that sufficient funds will remain in the account to pay a bill of £100 per annum at present-day values for a period of five years.

For example, a replacement heater costs £1500 to purchase and £100 per annum to maintain for the life of the installation, which is four years. What is the installation's NPV, if interest is received at 11% and inflation is estimated at 6%?

$$ndr = (1.11/1.06) - 1 = 4.7\%$$

$$NPV = \text{Year } 0 + \text{Year } 1 + \text{Year } 2 + \text{Year } 3 + \text{Year } 4$$

$$= C_0 + \frac{C_1}{(1+r)} + \frac{C_2}{(1+r)(1+r)} = \frac{C_3}{(1+r)^3} + \frac{C_4}{(1+r)^4} \qquad (1.7)$$

$$= 1500 + \frac{100}{1.047} + \frac{100}{(1.047)^2} + \frac{100}{(1.047)^3} + \frac{100}{1.047^4}$$

$$= £1857.03$$

This method of adjusting future year's costs to allow for interest and inflation is called discounting. The technique is called discounted cash flow (DCF). Annuity tables detailing solutions for $1/(1 + r)$ are available to help ease the manual calculations.

By arranging this formula into tabular forms, variables to a cash stream may be easily added or subtracted (see Table 1.3). For example, the initial cash expenditure may be modified to account for capital allowances, and the following year's also if a writing-down allowance is applicable. In the third year, a major service may be due, increasing third-year costs; an allowance for salvage or resale may be incorporated. Also some allowance, such as a sinking fund, may be added to cover emergency breakdowns. Computer software providing tabulated results in this form is widely available for solving these types of cash-flow problems.

Table 1.3 Tablulated solution to an example

Year	Expenditure	Tax	Nett cash	$1/(1 + r)$	Present value
0	– 1500	+ 131	– 1369	1	– 1369
1	– 100	+ 35	– 65	0.955	– 62
2	– 100	+ 35	– 65	0.912	– 59
3	– 500	+ 175	– 325	0·871	– 283
4	– 100	+ 35	260	0.832	+ 216
	+ 500	– 175			
	Total NPV				1557.00

Assuming:
   (i) £1500 subject to 25% allowance on 35% corporation tax, all taken in first year.
   (ii) Major service in the third year at £500.
   (iii) Inflation 6%, interest 11%.
   (iv) Sale of heater in the fourth year for £500.
   (v) 100% corporation tax allowance on maintenance costs.
   (vi) Expenditure is designated as a negative sum to demonstrate funds withdrawn, income is therefore positive.

The use of the discounting method for financial evaluation of competing solutions will provide a list of various solutions, each with a corresponding NPV, similar to that shown below.

Project	NPV (£M)	Capital cost (£M)
A	28	11
B	27	9
C	30	11
D	25	10

The lowest overall-cost NPV option will be project D and would normally be selected. However, NPV does not automatically determine the solution which makes money work the hardest. Is the extra £1 million for scheme D worth the £2 million saving in NPV offered over scheme B? The schemes can therefore be further evaluated by ranking with a profitability index by using the formula:

$$P_i = NPV/C \qquad (1.8)$$

where P_i = profitability index, C = capital cost and NPV = next present value.

Calculating the project for the profitability index:

Project	P_i
A	2.54
B	3.0
C	2.73
D	2.5

Ranking the projects using the above formula, scheme B is the most profitable and should be selected.

The use of NPV for a single scheme will not mean much; the client is unlikely to invest such a sum today to cover his future costs, but will pay future resource costs out of future earnings. NPV is, above all else, a comparator between competing schemes that gives the decision taker a financial decision to set alongside other ranking decisions. Other decisions may of course be more subjective, such as:

Prestige: the impression the project gives to the corporate image.
Futures: potential future changes and fall-back plans.
Longevity: the intended life-span of the project.

The method can only be used as a tool in the design process to evaluate the financial differences between a finite number of mutually exclusive schemes.

Development of NPV will result in a series of curves depicting NPV against time as shown in Figure 1.1. This will facilitate the selection of the most suitable scheme for the design life of the project. The selection of the longevity of a system is problematic. The client may not be able to define the precise life of a project, and designing to an actual life is also complex. Consider a shop unit where the retail trade expects a maximum life of seven years. Should thinner materials be installed to a design life of seven years? No, because the decision hinges on the quality of finish needed to portray an image. Similarly, how long is the design life of a church? Will the windows last 150 years? It

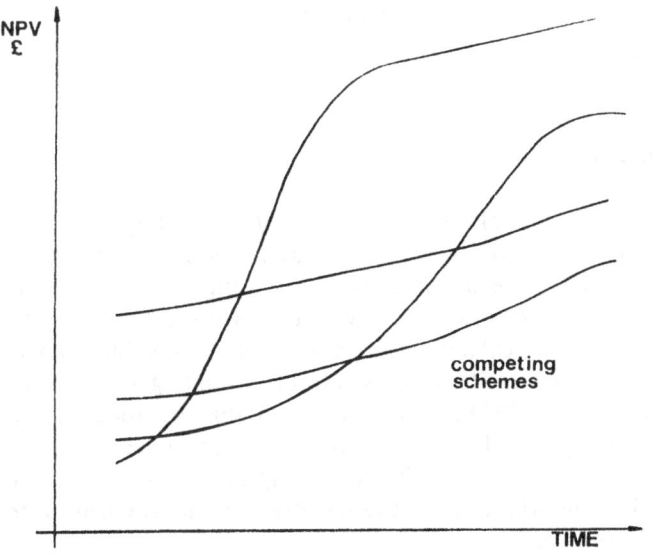

Figure 1.1 NPV against time for competing schemes.

is more likely that the client will have a range or period of years as being the actual life. NPV/time curves provide a visual representation of the financial performance of competing options over a range of years.

1.3.3   Internal rate of return

While simple payback and NPV are the most frequently used methods of evaluation in the building industry, a review of another commonly used method is also worthwhile.

The internal rate of return (IRR) is a DCF technique used where investment produces a return on capital employed. In using IRR, the capital cost is balanced against income to obtain a NPV of zero. The discount rate necessary is the IRR. This can be evaluated against an expected target for return on capital employed and the project's viability can thus be assessed, primarily against the expected performance of the business.

It could be considered that the need for comparison schemes is therefore eliminated, but in reality there is always a comparative scheme, i.e. the 'do nothing' scheme, against which IRR will be compared. This method, while used, does present difficulties in the construction context, in that it assumes that an investment will generate an income. It also assumes re-investment at the IRR. In construction terms, all cash flows are outwards and there are no guarantees of stable reinvestment levels.

The ultimate reason for using life cycle costing in the building industry has to be an auditable decision-making process to eliminate competing, mutually exclusive, schemes and not as a system from which profit can be made on its own merits. Profit determined by construction promoters will encompass a range of schemes at different sites with competing incomes. IRR can more profitably be utilised at this level rather than with individual schemes.

1.4   Sensitivity analysis

Applying a financial analysis to various solutions highlights a major problem with life cycle costing. While the mathematical model is rational and provides an auditable result, the uncertainty surrounding the variables used in the model throws a question over the viability of the solution. Systems of risk evaluation have been calculated in which a range of values of a variable are introduced and the sensitivity of a solution to changes in specific variables is assessed (Wright, 1973). If it is assumed that inflation in year 5 could vary between 5% and 10%, calculations can be made using a number of inflation rates for this year producing a range of results from which a decision can be made. This will indicate the sensitivity of the calculation to a change in a major variable.

Where multiple variables each have a range of possible results a method known as the Monte Carlo technique is used. This method utilises a series of random numbers for each variable, selects random solutions for each variable and calculates the result. Numerous calculations can then establish the probability of a final solution.

Mathematicians, economists and academics have therefore refined the computational techniques necessary to eliminate the majority of questions concerning the accuracy of life cycle costing. Indeed, computer-software producers have suitable proprietary programs in their portfolio that eliminate the need for the laborious calculations typical of a life cycle cost analysis.

1.5   Selection of the nett of inflation discount rate

The actual use of the information gained, its usefulness to the builder and its potential for incorporation into a construction design decision are other matters. It is doubtful whether any practising architect, engineer or quantity surveyor has embarked on a full analysis of options, based on the life cycle method, that has finally decided the building outcome, particularly in the private sector. Is this wrong? Should we consider seriously the use of life cycle costing for all buildings? In the USA, legislation brought in by the Carter administration, the National Energy Conservation Policy Act, makes life cycle analysis a legal requirement for all projects over $50 000.

This policy was a reaction to the energy crisis of the 1970s and numerous publications have appeared detailing how low capital-cost selections of projects have had disastrous resource-cost implications, particularly in the US defence industry. Should others follow this lead? An analysis of the results from a life cycle costing equation is worth inspection.

Figure 1.2 shows the results from an analysis of two buildings. Building A has a cheaper initial capital cost but gradually becomes more expensive over building B in terms of NPV as the life of the project continues. The centre line dictates the optimum solution in each case and is bordered by a band which represents the sensitivity of the calculation to variables. This band broadens with time as unknowns impact on the solution. It can be seen that three distinct areas are formed: two areas where it is obvious which building should be built, and a third, middle or 'indeterminate' period where the ranges of variables are such that either A or B could be said to be viable within the limits of a sensitivity analysis using assumed multiples of variables.

The slope of the line is dictated by the size of the discount factor applied. A low discount factor results in a steeply sloping line. A high discount factor results in a less steep line, extending the timescale during which the calculated balance point is achieved and increasing the 'indeterminate' period, as shown in Figure 1.3.

It is worth considering the discount factor in some depth. The selection of the ndr will significantly affect the design decision. An arbitrary selection

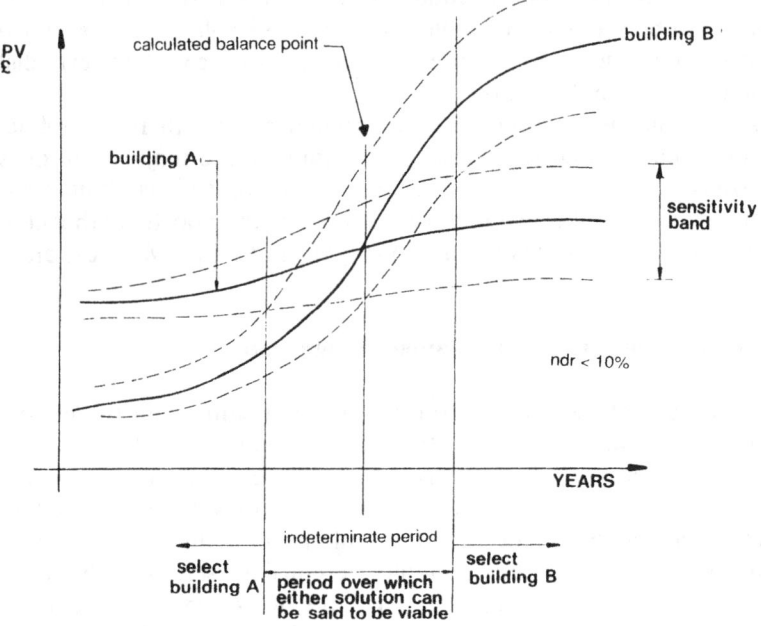

Figure 1.2   Analysis of two buildings, A and B, using a low discount factor.

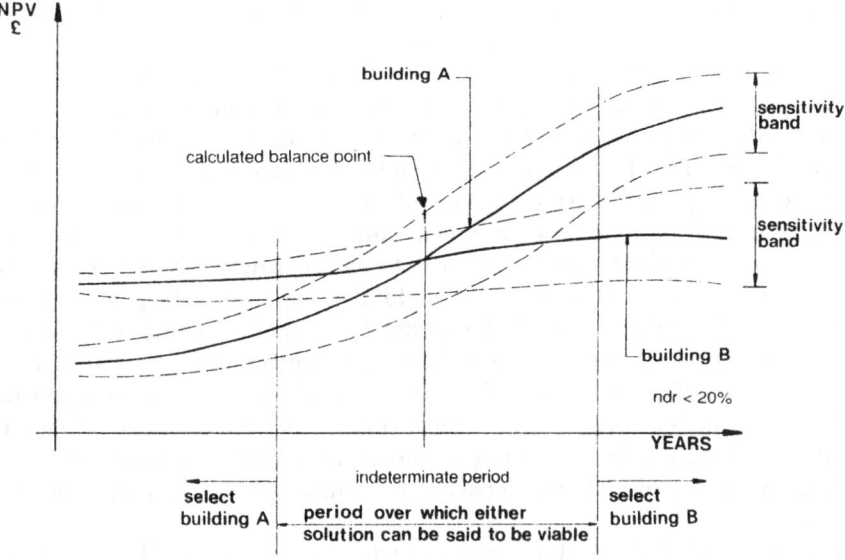

Figure 1.3   Analysis of two buildings, A and B using a high discount factor.

of a high rate will have a marked influence on the ultimate decision. Most texts on the subject suggest a discount factor of between 7% and 10%, with a nett of inflation rate between $2\frac{1}{2}\%$ and 5%. Such rates are based on values that can be demonstrated from secure fixed-interest investments such as Treasury Bonds, gilts or local interest rates. For a Treasury Bond rate of return of 9% and inflation at 6% the discount rate will be

$$ndr = [(1 + 0.09)/(1 + 0.06)] - 1 = 2.8\%$$

In practice, the use of this level of discount factor for public buildings is the sensible approach. In the private sector, however, investment in buildings is seen as a method of improving the return on investment in excess of secure investments; otherwise, building for profit will have little value. Most banks will not invest in a speculative venture of any kind unless it can be demonstrated that the return on the investment is in excess of 30%, thereby ensuring not only their own required return (usually 1–2% above LIBOR, the UK bank base rate) but also a level of return for investors such that they are unlikely simply to reinvest the loan elsewhere at a higher level of interest rate. Similarly, most developers will insist upon a developer's profit in the order of 20% for a building deal.

In private developments, therefore, which are often funded from simple debt financing with a proportion of equity investment, the required rate of return on the investment needs to be much higher, in the range 20–40%. If the required rate of return for a project is 35% and inflation is believed to be 6%, the nett of inflation discount rate will be

$$ndr = [(1 + 0.35)/(1 + 0.06)] - 1 = 27\%$$

It is obvious from Figure 1.3 that with high discount factors the indeterminate period is extended. A view will be taken that it is sensible not to invest additional funds now which could be invested elsewhere. When the return is at such a distant drop the lowest-cost selection system will tend to prevail on high discount-factor projects.

Does this statement therefore invalidate the use of life cycle costing in construction terms? Another factor that has an impact on the slope of the curve is the ratio of resource or running costs to the capital cost. Table 1.4 lists capital-cost items and Table 1.5 running-cost items. If the total annual

Table 1.4 Building: capital-cost items

Land costs
Professional fees (architect, engineer, lawyer)
Constructional costs
Commissioning costs
Promotional and sale costs
Funding costs
Management costs

Table 1.5 Building: resource-cost items

Furnishings
Routine maintenance
Servicing
Cleaning
Management
Energy costs
Rates and taxes
Sewerage
Breakdown repair/replacement
Salvage
Funding costs

resource costs are high, relative to the initial capital costs, the NPV line slopes steeply and high resource-cost projects will not have a high viability with ageing.

As important as discount factor and the resource-to-capital ratio is the selection of project longevity. Many projects now have a design life which is easily incorporated into the life cycle costing calculation. In practice, unquantifiable variables introduced during the project lifetime will inevitably lengthen or foreshorten the project life making the life cycle costing calculation potentially invalid. It has been suggested that a system of continuously updating the calculation by a system of life cycle cost management should be used (Flannagan and Norman, 1989). The updating of the life cycle costing equation throughout the life of a project has considerable merit; the introduction of improvements or new technology at the wrong time into a project could prevent a project's investment funds working at their hardest. In practice, it is doubtful if sufficient funds or management time will be devoted to this idea, however laudable the aims.

In most cases, any improvements not undertaken as part of a major refurbishment will disrupt a carefully calculated life cycle costing analysis and in principle should be discouraged, unless a significant improvement in returns can be demonstrated (e.g. less than twelve months on an adjusted simple payback calculation).

1.6   The application of life cycle costing

Life cycle costing (LCC) is a fundamental part of a decision-making process such as that illustrated in Figure 1.4. This process can be explained as an argument, which establishes a proposition, from which one deliberates upon alternatives, in order to decide upon a course of action.

The argument will define certain needs, the proposition will establish goals from which methods are developed to achieve the goals. Eventually a decision will be made and a solution reached.

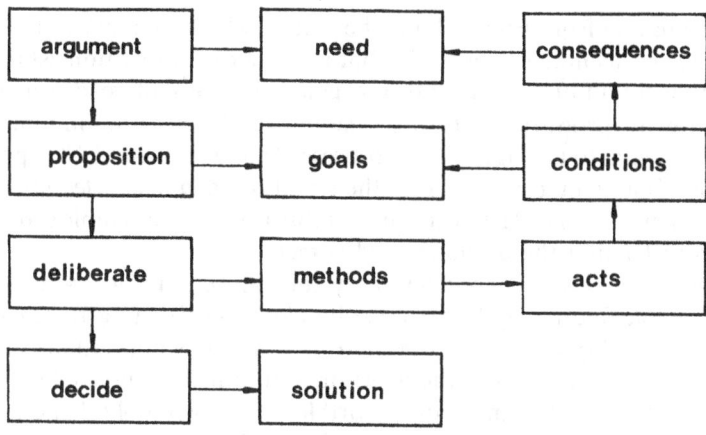

Figure 1.4 The decision process.

The methods deliberated upon could be described as acts that will be determined by conditions which must achieve certain goals. The conditions and acts will have consequences which must meet the needs established by the argument.

Consider a business producing paper clips; the business goal is to make profit from the production of paper clips. The decision process surrounding the business follows the logic path as shown in Figure 1.5. If a decision taken under deliberation affects the profit from sales, the ultimate need, i.e. to make money, will be similarly affected. This process is simple and obvious to understand. However, when the deliberation act is deferred or subcontracted to a third party, say an architect or builder, some of the original argument may be lost.

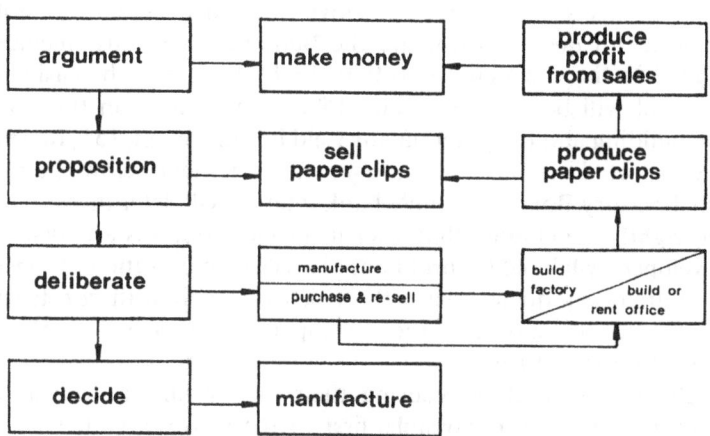

Figure 1.5 An example of the decision process.

What is missing from Figure 1.5 is the rate at which money can be made; it will always be 'as much as possible', but in practice competition establishes a reasonable minimum. Invariably the goals box should read 'sell paper clips in order to produce a return on capital of 30% as a minimum'. This information is vital to any agent in the process of conducting an LCC appraisal of any individual item, right down to the structural engineer selecting a steel beam for a new factory. The criterion of return on capital employed has to be used in order that the ultimate need is met.

For reasons of business confidentiality this figure may not be available, but a reasonable figure may be assumed by parties conducting LCC calculations. Confidentiality is necessary because this type of information is fundamental to the business plan of the investor. Without this knowledge the use of LCC as a decision-making tool is worthless. This is amply demonstrated by the use of simple payback without knowledge of the taxation implications.

In developing the mathematical model and identifying the parts played by the many variables applicable to the subject, we can see that before anyone can use a financial appraisal to select between options, it is essential that an intimate knowledge of the goals and ambitions (i.e. the propositions formed by the building promoter) is divulged to the design team. In order to do this, it is necessary to understand the basic propositions formulated by building promoters and decide how an LCC analysis may assist their ambitions.

1.6.1   The developer

The developer's decision-making process is controlled by the simple formula:

$$\text{Yield} = \left[\frac{\text{Income} - \text{resource costs}}{\text{Capital expenditure}} \right] \times 100\% \qquad (1.9)$$

In simple terms, this is the return on the investment in a project. Even if the developers have no desire to operate the building, they must include an attractive yield in the calculations so that the building may be easily sold. The target yield will be between 4 and 15% dependent upon the residual val ie of the building and the potential for yield improvement; 15% represents the average expected on an equity investment. Any yield below the rate of return from Treasury Bonds or simple bank interest will deter investors from the project, rightly concluding that safer investment options are available.

Most developers will build for profit out of a sale on completion of construction. Their purchasers do not want the risk associated with construction. For taking the construction risk the developer expects a high profit, 20% being a typical target return.

The developers' method of securing their high return is to make the development as attractive as possible, first to the prospective tenants who will provide the income, and secondly, to the investors who will wish to let the building early so that a purchaser will be attracted by a secured income.

In simple terms the formula will operate as follows:

Construction cost	£30 M
Developers' profit 20%	6 M
Sale price	£36 M

Yield required by purchaser 12%

$$\text{Income required from tenant} = (12/100) \times £36\,M + \text{resource costs}$$

$$= 4.32\,M + \text{resource costs}$$

With the tenant paying for the resource costs under a fully repairing lease (a lease that lays all the resource costs at the tenant's door), the resource costs to the purchaser are limited to management and debt recovery. In the example these may amount to £60 000/annum. Therefore the required income from a tenant = £4.38 M per annum. If the nett lettable area is, say, 100 000 ft², the charge to the tenant will be £43.8 per ft² per annum.

The factors that will influence a tenant's decision to enter into a lease will be many, and the priority of the factors will vary from tenant to tenant, major factors being:

- Location
- Facilities
- Prestige
- Terms of lease (cost, review periods)
- Running costs

In this environment what basis is used to formulate the parameters for an LCC appraisal? Is the installation viewed from the developer's or the tenant's point of view? At the time of construction the tenant's tax situation or expected return on capital employed is unknown. The developers, on the other hand, have no resource costs for the evaluation. They have no requirement for an LCC appraisal unless they can raise the rent rate because they can prove to a prospective tenant that the running costs for the building are less than other buildings, therefore yielding a better deal for the tenant. For example, if annual resource costs are 1/20 of capital costs, by implementing measures that result in savings in resource costs of 15%, the potential saving in the rent rate will be:

$$\text{Resource costs} = 30 \times 1/20 = 1.5\,M \text{ per annum}$$

$$\text{Saving generated} = £1.5\,M \times 15\%$$

$$= £225\,000$$

$$\text{Saving per ft}^2 = \left[\frac{£225\,000}{100\,000\,\text{ft}^2} \right] = £2.25/\text{ft}^2$$

The developer could therefore increase the rent by (2.25/43.8) or 5%. This has the result of increasing the yield to 12.77%. However, the tenant will need to feel sure that this resource saving is secure and not some paper figure. In this event the developers will have to satisfy the tenant of the cost of resources. This could tie the developers into a product-performance guarantee which they have no desire or legal obligation to do.

The potential for providing cost saving features of 15% resulting in an increase in rent of 5% does little for the developers. The figure is so small that it is likely to be lost simply in negotiations with the tenant. So why bother? Reduced running costs will be used by the developer as a 'shop window' or 'come on in and look' approach. This 'perceived' saving may attract the energy-conscious or 'green' tenant. No guarantees are offered other than a paper performance. The tenants' decision will be based upon a belief that they are leasing a lower running-cost building. The classic example of this is the debate between electric and gas central heating.

If the building is suitably insulated the running cost of electric storage heating is similar to that of gas heating, but the initial capital cost for electric storage heating is smaller. LCC techniques have been used to confirm electric storage heating as being the ideal option (Grahame, 1981). Yet gas central heating is universally considered to be cheaper to run. The reasons are that gas is perceived to be cheaper and that people prefer radiators to storage heaters, and buildings with gas central heating are therefore let more easily. The only decision for gas over electric is the attractiveness of the system to the tenant; LCC plays no part in the decision process for the lessor or the lessee.

Returning to the original theme, the original goals established in the proposition must be understood before the options are evaluated. The developers' goals are:

- Make money fast
- Buy low
- Sell high
- Attract potenial tenants
- Increase yields

The engineer faced with several options for heating, therefore, should not use LCC as a method of supporting the decision. Decisions on the specification of the building will be based not on cost analysis but on maintaining the cost of the building, maintaining a competitive rental, and yet still attracting a tenant. These are decisions that can only be subjective and can be quantified only against other lettable buildings in the area.

1.6.2 The institutional investor

The institutional investor is a building promoter who is cash-rich and seeks to invest for a return in excess of bank interest rates. Typical of this sort of investor would be a pension fund. Pension funds will spread risk across various market sectors and have a percentage of their funds invested in a property portfolio.

These investors will know which properties return good yields and in many cases will specify the standards of buildings they are willing to purchase from a developer. While owning a building they will have little interest in the operation of the building. This is facilitated by the 'fully repairing lease' whereby the tenant agrees to repair and maintain the building and its contents throughout the life of the lease. In this way the relationship of the resource costs to the capital cost are divorced and the germaneness of LCC is lost to the designer.

The institutional investor's goals are:

- High yield
- High retained value of building
- Low management cost
- Secure and happy tenants
- 'Blue chip' tenants (e.g. ICI, Shell, etc.)
- Long leases
- Frequent upwards-only rent reviews

Applying a LCC analysis to a building for most institutional investors will be a waste of time and effort. The selection of parameters such as discount factor for design decisions being a hopeless task, the designer will have no knowledge of the building user's required return on capital employed; indeed the designers may never get the opportunity even to know the name of the pension fund.

However, the use of DCF for the investor's funds is highly relevant, but only when applied to the cash stream and not to design decisions on the building construction.

1.6.3   The business

Most businesses will generally choose to lease a building, for various reasons such as taxation, cash flow, etc., but those businesses which choose to build for themselves will have a use for LCC.

If the designers should decide that an economic argument for a decision between options is relevant they must establish the following:

1. The business's expected return on capital employed for use as the interest rate.
2. The expected longevity of the building.
3. The way in which the resources are to be managed in order that costs can be established.
4. The business's predictions for inflation.

Most businesses that build buildings have no interest in profiting from the building itself; the building is a means to an end. The inevitable application of a high return on capital employed will predetermine the use of the cheapest solution in all but the systems that have the most heavy use of resources—for example, when the resource costs per annum are more than one-third of the capital cost. Individual LCC appraisals of high resource-cost items such as high energy users (e.g. windows, roofs, boilers, chillers, etc.) will have relevance, and should be undertaken as individual exercises to compare options between competing individual building components rather than for the building as a whole.

It is interesting to note how many businesses use DCF in their own investment decisions. Many operations are based on very short-term investment appraisals, and DCF plays little or no part. Simply asking the company accountant if DCF is used in the investment decision process will highlight the relevance of LCC for the project.

In the business environment little sympathy will be found for extra capital funds to reduce costs in years 10, 11 and 12. The entrepreneur will want to release funds now for injection into the mainstream business.

The goals for the business are:

● Low cost
● Prestigious building
● Locality
● Functionality

1.6.4   The public sector

This sector is one area where LCC has significant relevance. The need to quantify and justify expenditure from the public purse makes LCC particularly relevant.

The discount factor to be applied in public buildings is an area where some debate is certain. Arbitrary figures, as suggested in the USA, of 10% hardly seem relevant. The use of the Treasury Bond rate or index-linked gilts would appear to be sensible. Once a government forecast for inflation is added, the net of inflation discount factor will reduce to around 5% and a full LCC analysis will be worthwhile.

The goals for public sector building promoters are:

● Building functionality
● An auditable decision making process
● Cost effectiveness

1.6.5   Other uses

Applying LCC accounting to individual items within buildings, as suggested above, can have interesting impacts upon building design. Since the 1930s the basic apparatus for building services engineering (air-conditioning, heating, ventilation, public health and electrical services) has changed little; steel ducts, water pipes and copper cables are still used for the transportation of energy.

Items that have high capital costs and low resource costs (i.e. gently sloping NPV curves) will have a large impact upon an economic decision for system selection when a lowest-cost option is to be adopted. Items having such economic qualities are the energy transporters such as cables, pipes and air ducts.

Items having a relatively high resource-cost to capital-cost ratio are usually the prime elements of the system such as boilers, air-handling units, electric motors and light fittings, which are fundamental to the system's operation. Logic has it that we must reduce the amount of relatively high-cost energy transporters to make our system more LCC efficient.

Decentralisation of major services plant will therefore be more LCC-efficient for the building process. Interestingly, it is the very installation of these transportation systems that slows down the building process during construction, delaying completion. Accommodation of this equipment within the building also makes the building larger and more expensive. LCC supports a decentralised system of building engineering-services installation which has many 'knock-on' benefits to the building process.

In this way, while LCC may not have any obvious attraction to building promoters, if applied to comparisons of small elements of a building using indicative discount factors, LCC is a useful tool in identifying areas of a building that could be tackled in a different way to that normally accepted in the construction process.

1.7 Value engineering

Applying LCC to buildings should therefore not be undertaken lightly. The worth of the probable results of such an analysis should be evaluated before the exercise is undertaken. The danger inherent in a full analysis is not in the analysis but in the likelihood of decisions being taken that are impossible to include in the calculation.

Subjective decision making such as 'I like the look of that building, it will enhance my company's corporate image' can destroy a complex and intricate LCC analysis. Recognising this, a new term has developed, 'value engineering'.

Value engineering seeks to define the goals set by a proposition on a broader front than a simple economic analysis. Values are attributed to both objective and subjective arguments and decisions taken. Value engineering is a logical extension of LCC, which in its turn now forms only part of a subject that has derived from it.

References

Flannagan, R. and Norman, G. (1989) *Life Cycle Costing for Construction*, Surveyors Publications.

Flannagan, R., Norman, G., Medows, J. and Robinson, G. (1989) *Life Cycle Costing, Theory and Practice*, B.S.P. Professional Books, Oxford.

Grahame, G.D. (1981) The energy factor, paper given at the 51st Annual Conference of the Institute of Baths and Recreational Management.

Wright, M.G. (1973) *Discounted Cash Flow*, McGraw-Hill.

Bibliography

CIBSE Guide (1986) Section 18, Owning and operating costs. Dale, S.J. (1990) Difficulties with life cycle costing, *Professional Engineering*, January 1990.

Dell'Isoca, P.E. and Kirk, S.J. (1991) *Life Cycle Costing for Design Engineers*, McGraw-Hill.

Jeffrey, R.C. (1975) *The Logic of Decision*, McGraw-Hill.

Leech, D.J. (1983) *Economics and Financial Studies for Engineers*, Ellis Horwood, Chichester.

2 Life cycle costing related to reliability-based and optimum design

K. KOYAMA

2.1 Introduction

Civil engineering structures or structural systems are usually constructed for use by the public and for the purpose of investment. It is, therefore, required that the structure be safe and serviceable throughout its life cycle. Furthermore, it must the economic.

The safety and serviceability of structures or structural systems has generally been evaluated using reliability theory. This has been the case for the last decade. This ensures that the safety of structures is based on a probabilistic approach. The measure of safety is, therefore, expressed either by the probability of failure P_f or alternatively by the reliability $1 - P_f$. Exact calculation of the probability of failure is performed by way of multi-integrals of joint probability density functions, of both member resistances and loads or load effects. The calculation is very difficult and sometimes impossible to achieve, because structures consist of many complex member elements which are loaded in many different ways.

In order to escape the difficulty of multi-integrals, simplified approximation techniques or assumptions are used. The main objective of estimating safety is to determine whether a certain structure will fail only outside certain specified minimum limits. The minimum possibility of failure specified for a structure should be decided and authorized based on the importance of the structure or on the risk that society is prepared to accept of the structure failing.

For serviceability, the failure conditions are the safety conditions but a difficulty exists in reliability-based serviceability design. It is considered that serviceability is a function of structural behaviour and thus it is not clearly defined what serviceable is. This makes the serviceability design difficult to obtain, unlike the solution to a safety design. Consequently, design using fuzzy theory is used as it is able to express the ambiguous or unclear meanings of serviceability design.

Meanwhile, the optimum design of a structure is conceptually a design scheme to study the economic aspects of structural design. Structural optimization problems are usually formulated as the minimizing of an objective function expressed by design variables, subject to constraints which include side constraints such as lower or upper limits of sectional areas. To obtain an optimum solution, linear or nonlinear programming techniques are used. Monte Carlo methods or dynamic programming techniques are also adopted depending on the problem to be solved. In this chapter reliability-based design to optimize the life cycle cost is studied and investigated.

2.2 The basic reliability theory

Assuming only a single failure mode, the so-called failure criterion which creates a failure surface is given as:

$$Z = g(X_1, X_2, ..., X_n) = 0 \qquad (2.1)$$

in which $X = (X_1, X_2, ..., X_n)^T$ denotes the vector of the random variables of resistances and loads.

Structural failure occurs when Z becomes less than zero. The probability of failure, P_f, is obtained as:

$$P_f = \int \cdots \int_{\{x|g(x) \leqslant 0\}} f(x_1, x_2, ..., x_n) dx_1\, dx_2 ... dx_n \qquad (2.2)$$

in which $f\{x_1, x_2, ..., x_n\}$ is a joint probability density function of x. The integration is performed over the area where $g(X) \leqslant 0$.

As a simple case, consider the failure criterion to be represented by only two independent variables:

$$Z = g(R, S) = R - S = 0 \qquad (2.3)$$

where R is the strength and S is the load or load effect with the same units as R. This means that if S has units of stress, then R must also be expressed in units of stress.

The probability of failure P_f in this case is given as follows and is as shown schematically in Figure 2.1.

$$P_f = Pr(Z \leqslant 0) = Pr(R - S \leqslant 0) = \int_0^\infty F_R(x) f_s(x) dx \qquad (2.4)$$

or alternatively

$$P_f = \int_0^\infty f_R(x)\{1 - F_S(x)\} dx \qquad (2.5)$$

in which $f_R(x)$ and $f_S(x)$ are probability density functions of R and S, and $F_R(x)$ and $F_S(x)$ are their distribution functions, respectively. When R and S are normally distributed random variables, the probability of failure is simple and is given exactly as:

$$P_f = \Phi(-\beta) \qquad (2.6)$$

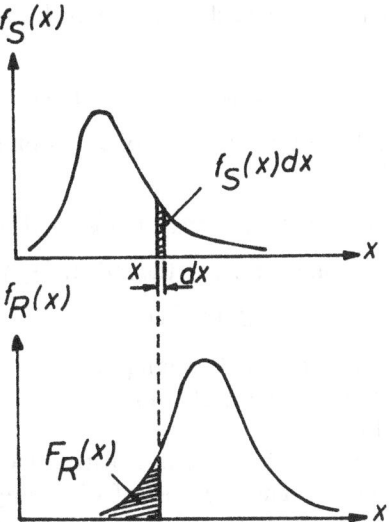

Figure 2.1   The probability of failure for R and S.

in which $\Phi(-\beta)$ is a standardized normal distribution function and β is so-called second moment safety index; β is shown as:

$$\beta = (\mu_R - \mu_S)/\sigma_{R-S} \qquad (2.7)$$

$$\sigma_{R-S} = (\sigma_R^2 + \sigma_S^2)^{\frac{1}{2}} \qquad (2.8)$$

$$\Phi(x) = \int_{-\infty}^{x} \phi(t)\mathrm{d}t, \qquad (2.9)$$

$$\phi(t) = \frac{1}{\sqrt{2\pi}}\exp(-\tfrac{1}{2}t^2) \qquad (2.10)$$

in which μ_R and μ_S are the means (first moment) and σ_R^2 and σ_S^2 are the variances (second moment) of R and S, respectively; β is called the second moment safety factor because it is expressed by using first and second moments only. The second moment safety index has a simple form and is a very good measure for estimating safety, related to probability, when R and S are normally distributed. It is, however, known that the safety index is not invariant to the expression of failure criterion and also not invariant to the distribution of R and S. This means that the result for β is not the same when the failure criterion is written as $R - S = 0$ and $(R/S) - 1 = 0$, although these two expressions have the same meaning as a criterion. The result for β is also not the same if either R or S is not normally distributed.

Until now, the invariant safety index β could be obtained, by definition, as the shortest distance from the origin to the failure surface in a reduced space. The reduced space is a transformed space from R and S to r and s and is shown as:

$$r = (R - \mu_R)/\sigma_R, \; s = (S - \mu_S)\sigma_S \qquad (2.11)$$

Using equation (2.11), the failure criterion of equation (2.3) can be rewritten as a function of r and s:

$$Z = g(r, s) = \sigma_R r - \sigma_S s + (\mu_R - \mu_S) = 0 \qquad (2.12)$$

From the definition of the safety index in the reduced space, β is expressed as:

$$\beta = (x^T x)^{\frac{1}{2}} \qquad (2.13)$$

in which $x = (r^*, s^*)^T$ is the so-called design point and is shown in Figure 2.2. It is also recognized that the result of equation (2.13) is equal to the safety index shown in equation (2.7).

Consider now that the variables are not distributed normally, nor is the failure criterion $g(x)$ linear. The method of obtaining the safety index is basically the same as that discussed above. Non-normal variates are approximated to normally distributed variates and the linearization of the failure criterion, using Taylor's expansion about an assumed design point X^0, is employed repeatedly until the safety index, which is defined as the shortest distance from the origin to the failure surface in reduced space, comes close to convergence. This can be shown briefly as follows. Linearization of the failure criterion:

$$Z = g(X) \doteq \nabla g(X^0)^T (X - X^0) = 0 \qquad (2.14)$$

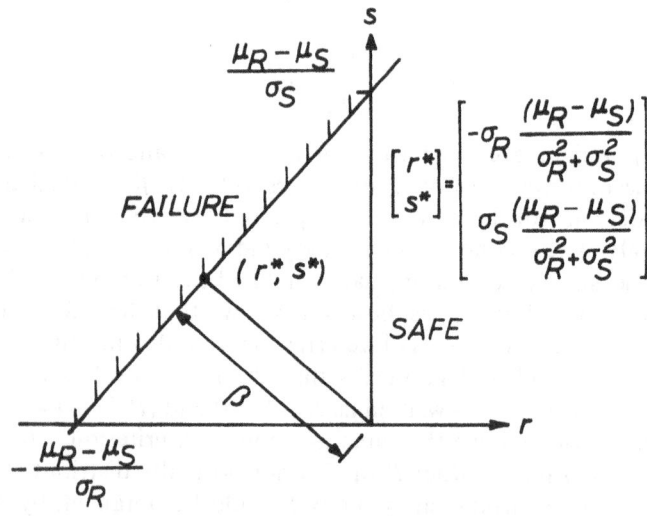

Figure 2.2 The safety index in reduced space in r and s.

transformation to a reduced variable:

$$x_i = (X_i - \mu_{x_i})/\sigma_{x_i} \tag{2.15}$$

approximation of a mean and a standard deviation (if X is a normal variate then the following equations are quite useless):

$$\mu_{x_i} = X_i^0 - \frac{\phi(\Phi^{-1}(F(X_i^0)))\Phi^{-1}(F(X_1^0))}{f(X_i^0)} \tag{2.16}$$

$$\sigma_{x_i} = \frac{\phi(\Phi^{-1}(F(X_i^0)))}{f(X_i^0)} \quad (i = 1, 2, \ldots, n) \tag{2.17}$$

Finding the design point:

$$x^0 = \nabla g(x^0)\left\{\frac{x^{0T}\nabla g(x^0) - g(x^0)}{\nabla g(x^0)^T \nabla g(x^0)}\right\} \tag{2.18}$$

Calculation of the safety index:

$$\beta = \frac{x^{0T}\nabla g(x^0)}{\{\nabla g(x^0)^T \nabla g(x^0)\}^{\frac{1}{2}}} \tag{2.19}$$

in which $f(\cdot)$ and $F(\cdot)$ are the non-normal probability density and distribution functions, respectively, and $\nabla = \partial/\partial x_i$ denotes the gradient. The calculation scheme, equations (2.14)–(2.19), is used repeatedly until the design point x^0 and the safety index, β come into good agreement.

The safety index used in reliability theory will be briefly discussed for simplicity. The more complex conditions, including when the variables or the failure modes are correlated mutually and the scheme to calculate β in the original space, are not discussed here. The basic concept and the calculating process described here, however, are useful enough for the problems encountered in civil engineering.

2.2.1 Example 1

Consider that the resistance R and load effect S are assumed to be normally distributed variables and that their means and variances are given as $R = N(13.72\,\text{KN/cm}^2, 1.372^2\,\text{KN/cm}^2)$, $S = N(7.84\,\text{KN/cm}^2, 1.568^2\,\text{KN/cm}^2)$. Calculate the probability of failure. Transform R and S into reduced space r and s as follows:

$$r = \frac{R - 13.72}{1.372}, \quad s = \frac{S - 7.84}{1.568}$$

Using equation (2.12), the failure criterion is shown as:

$$Z = 1.372\,r - 1.568\,s + (13.72 - 7.84)$$

The safety index is therefore obtained as:

$$\beta = \frac{13.72 - 7.84}{(1.372^2 + 1.568^2)^{\frac{1}{2}}} = 2.82$$

The probability of failure is found, using the table of cumulative standard normal distribution, to be:

$$P_f = \Phi(-2.82) = 0.0024$$

2.2.2   Example 2

Consider now that R and S are log-normally distributed and that their means and variances are given as $R = LN(\mu_{\ln(R)}, \sigma^2_{\ln(R)})$, $S = LN(\mu_{\ln(S)}, \sigma^2_{\ln(S)})$. Determine the safety index.

R and S are normally distributed by taking the natural logarithms as $\ln(R)$ and $\ln(S)$. In this case, the means the variances of $\ln(R)$ and $\ln(S)$ are obtained as:

$$\mu_{\ln(R)} = \ln(\mu_R) - \ln(1 + V_R^2)^{\frac{1}{2}}, \quad \mu_{\ln(S)} = \ln(\mu_S) - \ln(1 + V_S^2)^{\frac{1}{2}}$$
$$\sigma^2_{\ln(R)} = \ln(1 + V_R^2)^{\frac{1}{2}}, \quad\quad\quad \sigma^2_{\ln(S)} = \ln(1 + V_S^2)^{\frac{1}{2}}$$

in which μ_R, μ_S and V_R, V_S are means and coefficients of variation of R and S, respectively.

The transformation is made as:

$$r = \left[\frac{\ln(R) - \mu_{\ln(R)}}{\sigma_{\ln(R)}} \right], \quad s = \left[\frac{\ln(S) - \mu_{\ln(S)}}{\sigma_{\ln(S)}} \right]$$

The failure criterion, therefore, is shown as:

$$Z = \ln(R) - \ln(S) = \sigma_{\ln(R)}r - \sigma_{\ln(S)}s + (\mu_{\ln(R)} - \mu_{\ln(S)}) = 0$$

The safety index is obtained as:

$$\beta = \frac{\ln(\mu_R/\mu_S)\{(1 + V_R^2)/(1 + V_S^2)\}^{\frac{1}{2}}}{\{\ln(1 + V_R^2)(1 + V_S^2)\}^{\frac{1}{2}}} \doteqdot \frac{\ln(\mu_R/\mu_S)}{\sqrt{V_R^2 + V_S^2}}$$

The approximation in the above equation holds when V_R and V_S are small, less than say 0.3.

2.3   Optimum design

The optimum designs of structures are generally formulated as:

$$\text{minimize } z = f(x) \tag{2.20}$$

$$\text{subject to } g_i(x) \leq b_i, \ (i = 1, 2, \ldots, m) \tag{2.21}$$

in which x is the design variable to be optimized and m is the number of constraints: $f(x)$ and $g_i(x)$ denote the objective and constraint functions including any side constraints such as upper or lower limits of sectional area.

The objective is usually represented as a monetary function, the total weight of structural members or something else, and the constraints are often stresses in members or deflections of some point in the structure.

The solution is obtained by performing linear or nonlinear programming problem-solving techniques, depending upon whether $f(x)$ and/or $g(x)$ are linear or nonlinear functions. If objective or constraint functions are nonlinear, they are linearized and a linear programming technique is applied repeatedly, to gain an optimum solution. The method is called sequence linear programming (SLP). Otherwise the problem is changed from a problem having constraints to one having no constraints. This is called the SUMT (sequential unconstrained minimization technique) method. This method needs no gradient of a function, unlike SLP. Sometimes, the method which uses Lagrange constants is useful for solving problems with equations that have a strict equality condition.

2.4   Reliability-based optimum design

In reliability-based optimum design, it is usual to express equations (2.20) and (2.21) as shown:

$$\text{minimize } z = \sum_{i=1}^{n} \rho_i A_i L_i \tag{2.22}$$

$$\text{subject to } \sum_{i=1}^{n} P_{fi} \leqq P_{fa} \tag{2.23}$$

in which ρ_i, A_i and L_i are the weight density, the sectional area and the length of member i, P_{fi} and P_{fa} are the probability of failure of member i and the overall probability of failure, as specified in the design. To obtain the solution to this problem, the failure criterion of each member should be expressed by the design variables to be optimized. The criteria have the same formulae as equation (2.1).

The design, however, does not take into account the life cycle cost of the structure. The life cycle cost of a structure is very important for estimating the cost optimization problem. The problem may, therefore, be changed to one of minimum total cost, and generally takes the form:

$$\text{minimize } E(C_T) = C_i + P_f C_f \tag{2.24}$$

in which $E(C_T)$ is the expected total cost during the life cycle, C_i is the initial or construction cost and C_f is the maintenance when the structure deteriorates, loses durability or serviceability and requires repair or improvement. P_f is the probability that such a condition will occur.

The expected total cost $E(C_T)$ is related to the probability P_f for the structure. Furthermore, P_f is a function of the safety index β as seen in equation (2.6).

The optimum solution of the problem is principally obtained by solving:

$$\frac{\partial E(C_T)}{\partial \beta} = \frac{\partial C_i}{\partial \beta} + C_f \left[\frac{\partial P_f}{\partial \beta} \right] = 0 \qquad (2.25)$$

To obtain a significant solution is difficult, as it requires an accurate estimate of the costs C_i and C_f. Assumptions are usually introduced to ease the problem and obtain a solution.

In this case, the constraint is the failure criterion expressed by equation (2.1). The safety index, β, at the design point, has to be optimized so that the total cost shown in equation (2.24) is minimized.

2.4.1   Example 3

Solve equation (2.25), with C_f constant and assuming the initial cost C_i is given as:

$$C_i = a\{1 + b\exp(\beta\sqrt{V_R^2 + V_S^2})\}$$

The solution is obtained by using equations (2.6) and (2.10):

$$\frac{\partial P_f}{\partial \beta} = -\phi(-\beta)$$

The safety index is therefore obtained as:

$$\beta = -\sqrt{V_\theta} + \sqrt{V_\theta + 2\ln\{C_f/(\sqrt{2\pi}ab\sqrt{V_\theta})\}}$$

in which

$$V_\theta = V_R^2 + V_S^2$$

2.5   Reliability-based optimal reinforcement cover thickness in a concrete slab for life cycle enhancement

The durability of concrete affects the estimation of the life cycle costs and the service life of a structure. Life is classified into three categories: (1) economic life, (2) serviceability life and (3) physical life.

The physical life of a concrete structure depends mainly on the durability of the concrete. Lack of durability causes cracks or spalling in a concrete member, which may reduce the serviceability of the overall structure and may well ruin the sense of aestheticism. Corrosion of the reinforcing steel will also occur when carbonation proceeds through the concrete cover thickness. It is well known that corrosion is affected by the concrete cover thickness and that the variability of cover thickness depends on the quality of the construction work. General concrete structures are therefore required to have a minimum cover thickness to maintain a strong bond between the steel and the concrete, to prevent corrosion, and to provide protection against fire.

It is shown in BS 5400 (1990) that the nominal cover thicknesses are based on the grade of concrete and the environmental conditions of where the concrete is to be placed. To secure the correct cover thickness is very important for structural durability.

The reliability-based optimum concrete cover thickness in a concrete slab with respect to life cycle costs is investigated in this section. The concept of minimum total cost is used, to obtain a solution that provides for the longest physical life before carbonation reaches the surface of the reinforcement and for the probability of corrosion of the steel to attain a certain level.

The expected total life cycle cost is shown by equation (2.24):

$$\text{minimize } E(C_T) = C_i + P_f C_f$$

In this problem, the failure criterion is given as:

$$Z = X_D - X_C = 0 \tag{2.26}$$

in which X_D = the cover thickness of the concrete and X_C = the carbonation depth from the concrete surface, as shown in Figure 2.3. The carbonation speed in concrete is affected by external factors, such as the environmental conditions in which the concrete structure is placed, and internal factors such as material qualities. It is known that the carbonation speed is proportional to the square root of T, the total life term of the structure. Many formulae for carbonation speed have been proposed, but, for simplicity, equation (2.27) is adopted here. The depth of carbonation of the concrete can be expressed as:

$$X_C = \alpha\delta\gamma\sqrt{T} \tag{2.27}$$

in which α is a factor depending on environmental circumstances, δ is a factor for the coating condition of the concrete surface and γ is a factor for concrete quality. The quality of the concrete is considered to be significantly

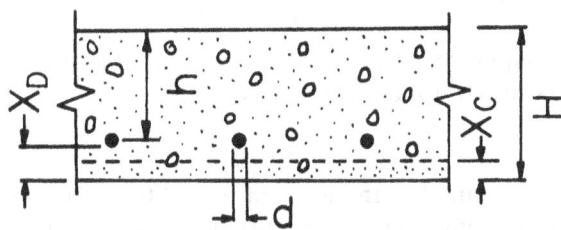

Figure 2.3   Covering thickness X_D and carbonation depth X_C in concrete slab.

affected by the water/cement ratio. Therefore, the formula for γ is given here as:

$$\gamma = \begin{cases} R(W/C - 0.25)/\sqrt{0.3(1.15 + 3.0\,W/C)}, & W/C \geq 0.6 \\ 0.37(4.6\,W/C - 1.76), & W/C \leq 0.6 \end{cases} \quad (2.28)$$

in which R is a factor depending on the quality of the concrete (here called the ratio of carbonation) and W/C is the water/cement ratio. This formulation is based on the proposal by the Japanese Architectural Standard Specification, JASS 5, *Reinforced Concrete Work*.

For general architectural structures that have a noncoated concrete surface and are placed outside, the factors α and δ in equation (2.27) and R in equation (2.28) are made equal to 1.0. It is assumed that the environmental conditions, the coating conditions and the concrete quality are the same for all civil engineering structures; therefore, these factors are taken as $\alpha = 1.0$, $\delta = 1.0$ and $R = 1.0$.

From the observed data, X_D and X_C are assumed to be normally distributed variables. The safety index for equation (2.26) is:

$$\beta = (\mu_{XD} - \mu_C)/(\sigma_{XD}^2 + \sigma_C^2)^{\frac{1}{2}} \quad (2.29)$$

in which μ_{XD} and μ_C are means and σ_{XD} and σ_C are standard deviations of X_D and X_C, respectively. It is also assumed that the relationship between the nominal value of X_D^n and μ_{XD} is:

$$X_D^n = (1.0 \sim 1.3)\mu_{XD} = f \cdot \mu_{XD} \quad (2.30)$$

Moreover, the standard deviations of X_D and X_C are given as:

$$\sigma_{XD} = \mu_{XD} V_D, \quad \sigma_C = \mu_C V_C \quad (2.31)$$

in which V_D and V_C are the coefficients of variations of X_D and X_C, respectively. The nominal cover depth of the concrete slab is therefore obtained as:

$$X_D^n = f \cdot \mu_C \cdot g(\beta) \quad (2.32)$$

in which

$$g(\beta) = \frac{1 + \beta(V_{XD}^2 + V_C^2 - \beta^2 V_{XD}^2 V_C^2)^{\frac{1}{2}}}{(1 - \beta^2 V_{XD}^2)} \quad (2.33)$$

Assuming α in equation (2.27) is a random variable, the mean of X_C is obtained by first-order approximation as:

$$\mu_C = \mu_\alpha \delta \gamma \sqrt{T} \quad (2.34)$$

in which μ_α is the mean of α, and is taken to be 1.0. The value is appropriate because general civil engineering structures such as concrete slabs are placed outside.

The initial cost in equation (2.24) is shown as:

$$C_i = W_s[A_s \cdot \text{costs} + \{(h + d/2) + X_D^n\} \cdot \text{costc}] \tag{2.35}$$

in which, A_s is the required reinforced area for unit length, h is the effective depth of the concrete slab, d is the diameter of the steel, W_d is the width of the slab, and costs and costc are the cost per unit volume of the steel and

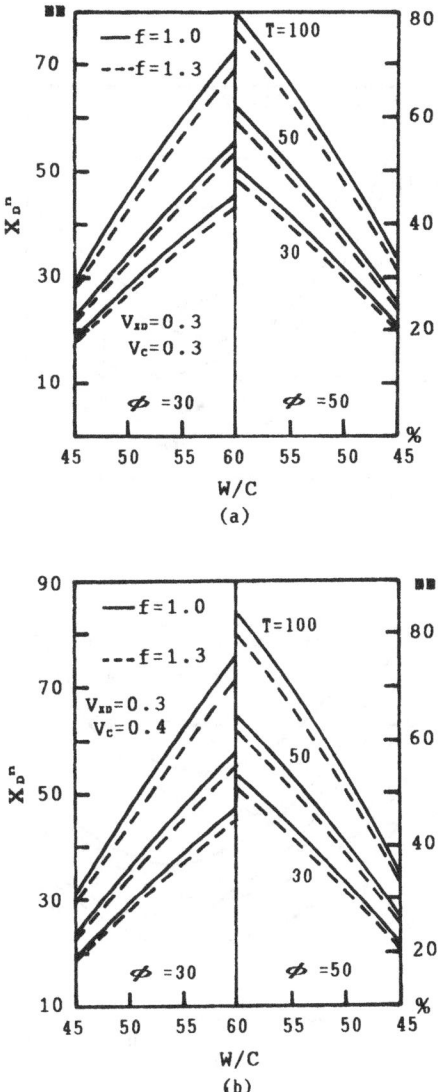

Figure 2.4 Required optimum cover thickness for water-to-cement ratio (W/C) and life term T.

the concrete, respectively. In these costs, only the material costs for the steel and concrete are included. Labour costs are not included. By using equations (2.32) and (2.35) and performing equation (2.25), the optimum safety index is obtained as:

$$\beta = \sqrt{2\ln(1/\text{WW})} \tag{2.36}$$

in which

$$\text{WW} = \sqrt{2\pi} \cdot f \cdot (1/\phi) \cdot \mu_C \cdot \partial g(\beta)/\partial \beta \tag{2.37}$$

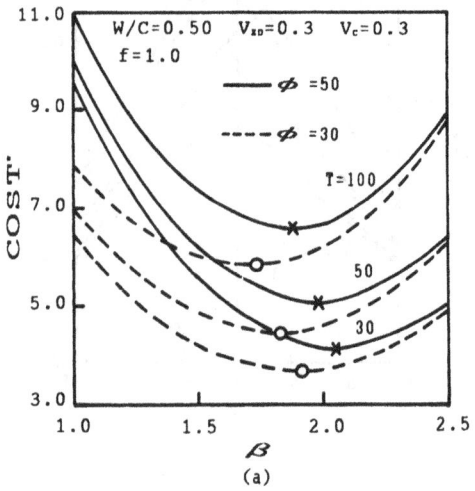

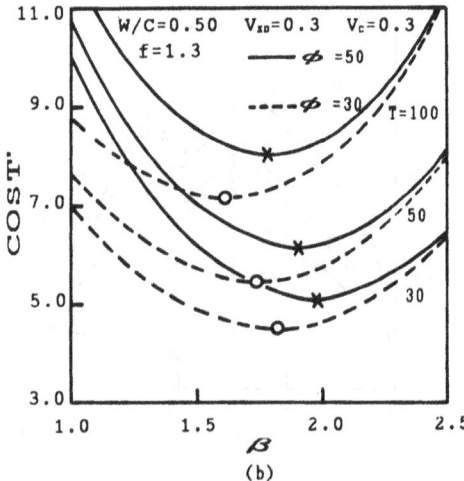

Figure 2.5   Expected total cost for safety index β.

and

$$\phi = C_f/(W_d \cdot \text{cost}c) \qquad (2.38)$$

It is recognized that the safety index β in equation (2.36) is not expressed in explicit form because β exists on both sides of the equation. Some mathematical technique is required to obtain a solution. For example, to obtain the optimal solution for cover thickness, the following assumptions are made. The life of the concrete slab varies from 30 to 100 years. The ratio of water to cement varies from 0.45 to 0.60. These are considered reasonable and acceptable assumptions for civil engineering structures. To estimate the cost C_f is difficult. Therefore, C_f is assumed to be related to construction cost, and is calculated in this section by assuming a cost factor ϕ in equation (2.38). The cost factor ϕ has the two values 30 and 50.

Solutions using these values are presented in Figures 2.4–2.6. The required optimum cover thicknesses for various water to cement (w/c) ratios are shown in Figure 2.4. In Figure 2.4a, the coefficients of variation V_{XD} and V_C of X_D and X_C are taken to be $V_{XD} = V_C = 0.30$. Figure 2.4b is for $V_{XD} = 0.30$ and $V_C = 0.40$. Accepting the constant part in equation (2.24), the dimensionless expected total cost, $\text{COST}^* = (f \cdot \mu_C \cdot g(\beta) + \phi \cdot P_f)$, is shown in Figure 2.5 for the safety index β, for $W/C = 0.50$, $V_{XD} = V_C = 0.30$ and $f = 1.0$ and $f = 1.3$. The white circles and the crosses in Figure 2.5 represent calculated COST^* based on the optimum safety index β_{opt}. Obviously, the expected total cost becomes a minimum at the optimum safety index β_{opt}. The solutions for β_{opt} are shown in Figure 2.6 for $W/C = 0.45$ and 0.55 and for various values of T.

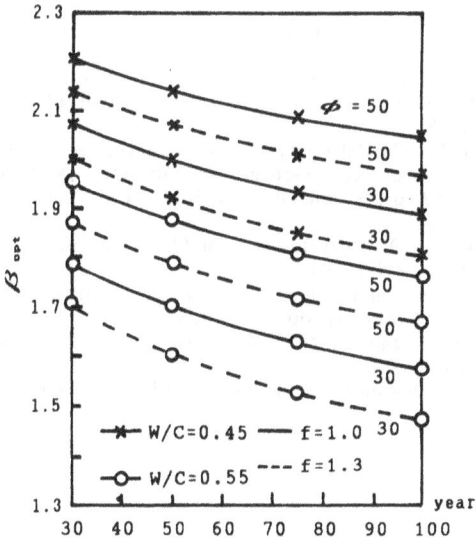

Figure 2.6 The optimal safety index β_{opt} for life term T.

2.6 Conclusion

In this chapter, reliability based design and optimum designs are used to optimize the whole life cost of a civil engineering structure. The form of the method is given, together with examples.

Bibliography

Ang, A.H-S. and Amin, M. (1969) Safety factors and probability in structural design, *ASCE*, No. ST7, 1389–1405.

Ang, A.H-S. and Cornell, C.A. (1974) Reliability bases of structural safety and design, *ASCE*, No. ST9, 1755 –1769.

Freudentahl, M., Garrelts, M. and Shinozuka, M. (1966) The analysis of structural safety, *ASCE*, No. ST1, 267–325.

Grigoriu, M. and Turkstra, C. (1978) Structural safety indices for repeated loads, *ASCE*, No. EM4, 829–843.

Hannus, M. (1973) Numerical analysis of structural reliability, Technical Research Centre of Finland, Building Technology and Community Development Publication 5, Helsinki.

Hasofer, A. M. and Lind, N. C. (1974) Exact and invariant second moment code format, *ASCE*, No. EM1, 111–121.

Hilton, H.H. and Feigen, M (1960) Minimum weight analysis based on structural reliability, *Journal of the Aerospace Sciences*, September, 641–652.

Japan Society of Architecture (1986) Japanese Architectural Standard Specification JASS 5, *Reinforced Concrete Work*, 7th edn (in Japanese).

Ken, K. (1989) Reliability based economic evaluation of structures considering the life term, *JSCE Structural Eng. Earthquake Eng.* 6(2), 357s–363s.

Ken, K., Hironobu, T. and Isao, Y. (1991) Reliability based optimal thickness of covering of concrete slab for lifecycle (in Japanese), in: *Proc. JCOSSAR '91: The Second Conference on Structural Safety and Reliability*, 221–228.

Lind, N.C. (1976) Approximate analysis and economics of structures, *ASCE*, No. ST6, 1177–1195.

Mau, S. and Sexthmith, R. (1972) Minimum expected cost optimization, *ASCE*, No. ST9, 2043–2058.

Moses, F. (1974) Reliability of structural systems, *ASCE*, No. ST9, 1813–1820.

Moses, F. and Kinser, D.E. (1967) Optimum structural design with failure probability constraint, *AIAA*, No. 6, 1152–1158.

Moses, F. and Stevenson, J. (1970) Reliability based structural design, *ASCE*, No. ST2, 221–244.

Parkinson, D. (1978) Solution for second moment reliability index, *ASCE*, No. EM5, 1267–1275.

Rackwitz, R. (1980) First order reliability methods, Technical University of Munich, Sonderforhungsbereich 96.

Rackwitz, R. and Fiessler, B. (1978) Structural reliability under combined random load sequence, *Computers and Structures*, 9, 489–494.

Switzky, H. (1967) Minimum weight design with structural reliability, in: *Proc. AIAA 5th Annual Structures and Materials Conference*, pp. 1152–1158.

Thoft-Christensen, P. and Yoshisada, M. (1986) *Application of Structural Systems Reliability Theory*, Springer–Verlag.

Veneziano, D. (1974) Contribution to second moment reliability theory, MIT Research Report No. R 74–33.

3 Life cycle costing related to the refurbishment of buildings

S.-I. GUSTAFSSON

3.1 Introduction

This chapter describes the use of life cycle costing when a building is to be retrofitted. The life cycle cost (LCC) includes all costs that emerge during the life of a building, such as building costs, maintenance costs and operating costs. When the LCC is to be calculated, future costs must be transferred to a base year by use of the present value method. Although the LCC includes all costs, this chapter will only consider those costs related to the heating of the building, or the use of energy in one form or another. Retrofits which allow a cheaper form of cleaning or result in a different aesthetic shape are not included. One other constraint is that all the consequences must be expressed in monetary terms. This chapter, however, deals with the implementation of extra insulation on various building parts, changing windows for a better thermal performance, weatherstripping, exhaust-air heat pumps and different types of heating equipment. The basic view is that the building is considered as an energy system and, at least sometimes, all the energy-conserving measures must be dealt with at the same time if an accurate result is to emerge. Another corner-stone of this chapter is that the retrofit strategy shall be the one with the lowest possible LCC, i.e. the situation must be optimized. Derivative, direct search and linear programming methods are dealt with and an extensive reference list is presented showing the state of the art in the middle of 1991. There are also many examples of real cases in order to highlight various aspects of this subject.

When a building is to be refurbished it is important to consider that it already has a life cycle cost (LCC) whether it is rebuilt or is left as it is. If the LCC is to be the ranking criterion for deciding what to do, it is important to compare the new LCC to the old, or existing, LCC. If the new LCC is lower, it is profitable to rebuild; if the opposite is true, the building should not be refurbished at all. One of the basic concepts in life cycle costing is the present value (PV) which is used for transferring future costs to one base year, where they can be compared properly. There are many papers and books concerning the use of the PV for life-cycle costing, e.g. Marshall (1989) and Flanagan *et al.* (1987, 1989), but only the expressions for calculating the PV will be shown here. The first expression, equation (3.1), shows the PV for a single cost occurring once in the future, while the second expression, equation (3.2), shows the PV for annually recurring costs:

$$PV_s = C_s \cdot (1 + r)^{-n} \tag{3.1}$$

$$PV_a = C_a \cdot [1 - (1 + r)^{-m}]/r \tag{3.2}$$

where r = the real discount rate, n = the number of years until the signal cost C_s occurs and m = the number of years in which the annual costs C_a occur.

Equation (3.1) is suitable for calculating the PV for, for example, window retrofits or insulation measures, while equation (3.2) is used for energy and other annually recurring costs. Before it is possible to start with the PV calculations it is necessary to find the costs C_s and C_a and proper values for r, n and m. Unfortunately, there are difficulties here, because of uncertainties both for the costs as well as for the economic factors. C_s might be found in certain price lists (see Gustafsson and Karlsson, 1988a, for an example of the calculation), so if these are accurate the problem is partially solved. C_a, however, is influenced by the thermal state of the building and large uncertainties due to the fluctuating energy price in the future. The real discount rate, r, cannot be set to an accurate value valid for all investors, and different authors recommend values between 3 and 11%. Van Dyke and Hu (1989) even show that some investors have dealt with negative rates. Note that inflation is excluded from these values. The value for n, the number of years until a retrofit is inevitable, can likewise not be predicted accurately and the same is valid for the projected life of the building, m. Many authors have dealt with this problem, and papers are frequently published in, for example, proceedings from CIB conferences.

From the above discussion it might seem hopeless to calculate anything at all and believe in the result. However, every time an investment is made, values for all the variables are set even if the investor is unconscious of them. A closer analysis will often reveal limits within which the values might move, and then it will be possible to calculate the result using different values for each calculation. Without computers this is a very tedious task and is one of the reasons why life cycle costing has not been frequently used. By using computers, large problems can be solved in a few minutes. It is nowadays possible to calculate the result for a number of different scenarios and then examine the situation in a so-called 'sensitivity analysis'. Several interesting results will then occur and general conclusions can be drawn in spite of the uncertainties in the input data.

3.2 Insulation measures

The optimal thickness of extra insulation is influenced by a number of variables; the building cost, the climatic conditions, the energy cost, etc. A suitable way to describe the building cost (BC) is as follows:

$$BC_{ins} = C_1 + C_2 + C_3 \cdot t_{ins} \qquad (3.3)$$

where C_1 = the amount in Swedish kroner (SEK) /m² for scaffolds, demolition, etc., C_2 = the amount in SEK/m² for the new insulation, studs, etc., C_3 = the amount in SEK/m² per metre for new insulation, studs, etc. and t_{ins} = the thickness of new insulation in metres (1 US $ = 6 SEK approx.).

The reason for splitting up the cost into three parts is the influence of the existing life of the building asset. As an example, consider an external wall. The facade is in a rather poor shape, but nonetheless, the retrofitting of it might not be necessary for, say, ten more years. The C_1 coefficient shows the amount of money to be paid at year 10 whether energy-conserving measures are taken or not. This retrofitting is called inevitable or unavoidable and is very important to take into consideration. Assume that C_1 equals 500 and that the wall must be retrofitted in year 10 when it is unavoidable. The real discount rate is set to 5%, while the project life is assumed to be 50 years. The life of the new facade is assumed to be 30 years. Consequently the PV of the retrofitting, by equation (3.1), will become:

$$500 \cdot (1 + 0.05)^{-10} + 500 \cdot (1 + 0.05)^{-40} - [(30 - 10)/30] \cdot 500 \cdot (1 + 0.05)^{-50}$$
$$= 349.0$$

This PV calculation shows the value of the money invested in year 10 and year $10 + 30$. Further, the salvage value at year 50 is subtracted. This PV must be added to the LCC of the existing building, because it shows the inevitable retrofit cost. If the wall is retrofitted now, at the present time, the PV calculation will become:

$$500 \cdot (1 + 0.05)^{-0} + 500 \cdot (1 + 0.05)^{-30} + [(30 - 20)130] \cdot 500 \cdot (1 + 0.05)^{-50}$$
$$= 601.21$$

From this it is shown that the increase of the cost for retrofitting now, instead of at year 10, is

$$601.2 - 349.0 = 252.2$$

The cost of 601.2 must thus be added to the new LCC. Closer details about PV calculations can be found in Ruegg and Petersen (1987).

After this, the cost for the insulation itself must be included. However, it is assumed that insulation is only installed once, at the base year, so it is not necessary to calculate the PV for the additional insulation. At this point in the examination, it is not possible to determine how much insulation is to be installed and subsequently included in the cost $C_3 \cdot t_{ins}$ in equation (3.3). It has been shown (Gustafsson, 1986) that the new U-value for an extra insulated asset may be expressed as:

$$U_{new} = U_{exi} \cdot k_{new}/(k_{new} + U_{exi} \cdot t_{ins}) \qquad (3.4)$$

where U_{exi} = the existing U-value in $Wm^{-2}K^{-1}$, and k_{exi} = the thermal conductivity in the extra insulation in $Wm^{-1}K^{-1}$.

Multiplying the U-value by, firstly, the area of a building asset, secondly, the number of degree hours for the building site and, thirdly, the energy price, will result in the annual cost for the energy flow through the asset. Further, the annual cost must be multiplied by the PV factor, calculated by using equation (3.2) which will yield the total energy cost for a number of years. Using a real discount rate of 5% and a project life of 50 years makes the PV factor equal to 18.26. In Malmö, in the south of Sweden, the number of degree hours for one year equals 114 008. It has been assumed that one degree hour is generated for each hour that the desired indoor temperature, 21°C, is higher than the outdoor temperature. (See Gustafsson (1986) for all details about degree hour calculations.) If the energy cost is 0.40 SEK/kWh, the area of the building asset 200 m², with an existing U-value of 0.8 and a k-value for the new insulation $0.04\,Wm^{-1}K^{-1}$, the cost (TC) in SEK for the energy flow through the building asset will become:

$$TC_{energy} = 114008 \cdot 0.40 \cdot 200 \cdot 0.8 \cdot 0.04 \cdot 10^{-3} \cdot 18.26/(0.04 + 0.8 \cdot t_{ins})$$
$$= 5329/(0.04 + 0.8 t_{ins})$$

When the building is extra insulated there is also a cost for the insulation and for correctly locating it. Assuming that the constant C_2 equals 100 SEK/m² and C_3 equals 600 SEK/m²/m, according to equation (3.3) the result for the building cost in SEK for the asset will be:

$$TC_{building} = 200 \cdot (601.2 + 100 + 600 \cdot t_{ins}) = 140\,240 + 120\,000 \cdot t_{ins}$$

The problem now is to minimize the sum of the energy and the building cost; this is done using the derivative of this sum which is set to zero. The way to do this is shown by Gustafsson (1986), but the result is that the optimal level of insulation in metres becomes:

$$t_{opt} = -(0.04/0.8) + (5329/120\,000 \cdot 0.8)^{0.5} = 0.186$$

Inserting this value for optimal level of insulation as t_{ins} in the equation above will result in a LCC current of 190 785 SEK. This cost is now to be compared to the LCC if the building is left as it is. For the current asset this is:

$$LCC_{exi} = 200 \cdot 349.0 + 5329/(0.04 + 0.8 \cdot 0) = 203\,025\ SEK$$

The existing LCC is thus higher than the new one. Even if the difference is as small as about 13 000 SEK, it is profitable to insulate the asset with the optimal amount of new insulation. In Figure 3.1 the situation is shown graphically. As can be seen, the existing LCC is higher than the optimal new LCC. If, however, the inevitable costs could be decreased, for example by assuming that the remaining life of the asset envelope is increased, the existing LCC will also decrease, and at a certain point it becomes better to leave the

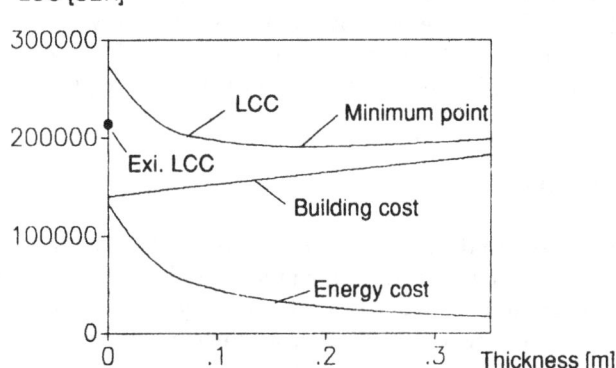

Figure 3.1 Graphic view of insulation optimization.

building as it is. From Figure 3.1 it is obvious that enough insulation must be applied. This limit is in the case above approximately 0.07 m. If less insulation is used, the retrofit will become unprofitable. If too much insulation is installed the same might happen, but if the graph is studied even in detail, this fact could not be observed. It is even better to use a 0.35 m thickness of insulation than not to insulate at all. In Gustafsson (1988) a thorough examination is made of all the parameters concerned.

3.3   Replacing windows

When the replacement of windows is being considered it is not easy to find a continuous function to derive in order to find the best solution, although there have been some attempts to find such a function (Markus, 1979). Instead, it has been shown that it is preferable to compare different sets of windows with each other. The existing LCC is thus compared with the new LCC for a number of alternatives. It is very important to find not just one solution with a lower LCC, but the lowest solution of all. It is also important to consider the fact that a thermally better window normally reflects solar radiation to a higher degree. This fact can be dealt with by use of a so-called shading factor. The situation will therefore differ for various orientations of the windows. The best solution may be, in the Northern hemisphere, to keep the double-glazed windows oriented to the south, while changing to triple-glazed windows to the north. Life cycle costing and windows are dealt with in more detail in Gustafsson and Karlsson (1991). The building cost for windows may be expressed as (Gustafsson, 1986):

$$BC_w = C_1 + C_2 \cdot A_w \tag{3.5}$$

where C_1 = a constant in SEK for each window, C_2 = a constant in SEK/m^2 for each window and A_w = the area in m^2 for one window.

Here, BC_w will appear whenever there is a change of a window and the expression is consequently used in a different way from equation (3.3).

3.4 Weatherstripping

Mostly it is profitable to decrease the ventilation flow through the building. This can be accomplished by caulking windows and doors. The cost for this measure is often very low compared to other energy retrofits, but nonetheless it is not always the best way to act, especially when exhaust-air heat pumps are part of the solution. It is also important to consider that it is necessary to ventilate the building; too much weatherstripping might make them unhealthy. When using life cycle costing, these facts are often hard to include in the calculations and only the energy costs are dealt with here. Suppose a building has fifty windows and doors to caulk. If the cost for caulking is 200 SEK/item the total cost will become 10 000 SEK. Further, assume that the weatherstripping must be repeated after 10 years. The PV cost will thus become approximately 23 600 SEK if a 5% discount rate and a 50-year project life are used. If the volume of the building is $5000\,\text{m}^3$ and the ventilation rate is 0.8 renewals per hour, the flow is $4800\,\text{m}^3/\text{h}$. The heat capacity for air is about $1.005\,\text{kJ}\,\text{kg}^{-1}\text{K}^{-1}$ and the density approximately $1.18\,\text{kg/m}^3$. Consequently the heat flow can be calculated to about $5\,700\,\text{kJ}\,\text{K}^{-1}\text{h}^{-1}$. If the same number of degree hours as before is assumed to be valid, i.e. 114 008, the energy flow will become 180.5 MWh/year. Using the PV factor 18.26 and an energy price of 0.4 SEK/kWh, as before, the total energy cost will become 451 000 SEK. If the ventilation flow is decreased to, say, 0.2 renewals per hour, this cost will become 338 000 SEK. It is obvious that weather-stripping, in this example, will be profitable.

3.5 Exhaust-air heat pump

One other means to decrease the heat flow from the ventilation is to install an exhaust-air heat pump. This device takes heat from the ventilation air and, by use of electricity, transfers the heat back into the building. One part of electricity may often result in two to three parts of heat. It is, however, very important to install a heat pump of the right size because the amount of heat in the ventilation air is a limited resource. In this chapter no example is presented of how to calculate the LCC for the heat pump. This is because it is very rarely chosen as an optimal retrofit. It must, nonetheless, be emphasized that using a heat pump might make it unprofitable to caulk the windows in the building. Even if weatherstripping is a very cheap retrofit, it might be even cheaper to use a slightly larger heat pump in order to utilize the increased ventilation flow from not caulking the windows.

3.6 Other building or installation retrofits

Exhaust-air heat exchangers are not dealt with here. This is because of the high cost of distributing the air from the exchangers to different apartments in a building, but the principle for the calculations is the same. Water-heater blankets and the regulation of radiator thermostats can be important measures in decreasing the energy need. However, the blankets are only useful if the water heater is located outside the thermal envelope or if the heating season is very short. Thermal thermostats will ensure that the desired inside temperature stays within defined limits, but they will only be useful if the surplus heat is wasted due to the use of extra ventilation.

3.7 Heating system retrofits

There are a number of heating system retrofits that must be considered. If the building is equipped with an oil boiler, it could be advisable to change it for one with a better efficiency. Or perhaps district heating would be preferable, if such a possibility exists. At least in Sweden, bivalent systems seems to be of interest when larger buildings are considered. A bivalent, or dual-fuel, system has an oil boiler taking care of the thermal peak load while a heat pump is used for the base load. It is important to optimize the sizes of the equipment and it has been shown that the correct level of extra insulation is, in this case, essential for reaching the lowest LCC (Gustafsson and Karlsson, 1988b). However, if the heating system is changed, it will lead to a building retrofit strategy that differs considerably from the one chosen when the original heating system is used. The process for calculation is depicted in Figure 3.2. Figure 3.2 also emphasizes that different retrofits might interact. Suppose an attic floor insulation was found to be profitable. When at the next retrofit perhaps additional external wall insulation is examined, the new LCC is compared to the original one, i.e. without additional attic floor insulation. Suppose also that this retrofit is profitable. The problem encountered is that, if the attic floor insulation is introduced, the external wall insulation might become unprofitable. Using an incremental method as above will overestimate the savings which are actually made. The method for optimization must consequently include an examination which includes the combination of the retrofits. If the difference between the incremental and the combination retrofit is very small the accuracy is satisfied; otherwise the insulation thickness must be changed and the resulting LCC must be recalculated. Perhaps the retrofit considered will drop away totally from the

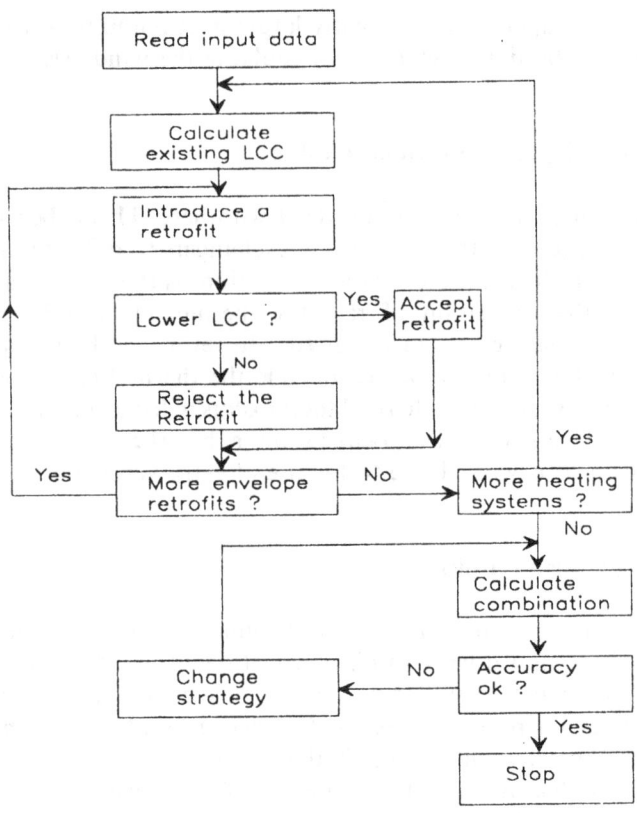

Figure 3.2   Optimization process (Gustafsson and Karlsson, 1989).

optimal solution. Fortunately, this interaction is usually very small, at least
if the best candidates for an optimal solution are examined. Sonderegger
et al. (1983) have calculated the difference to be about 2% in some cases and
usually the interaction can be neglected. It must be noted that sometimes
the situation is the opposite, i.e. interaction leads to a lower LCC for the
combination than for the incremental method. This has been observed for
fenestration measures and is discussed in detail in Gustafsson and Karlsson
(1991), but the cases where this fact has been observed are rare and probably
of academic interest only. In Table 3.1, a case study is presented clarifying
the above discussion. The original LCC is calculated to 1.48 million SEK.
The computer program used then checks to see if the attic floor insulation
is profitable. This is not the case and thus the value '.00' is shown on the line
below. External wall insulation, however, was found to be optimal and the
amount saved calculated to be 0.05 million SEK for the project life of the
building. Triple-glazing and weatherstripping were also candidates in the
optimal solution. If the existing heating system is changed to a new oil
boiler, the LCC is increased, even if the money saved by retrofitting is

Table 3.1 LCC table from the OPERA model (values in million SEK) (from Gustafsson, 1990)

	Exis. syst.	New oil	Ele. heat	Dist. heat	Gr. w. heat	Nat. gas	Tou. dist	Tou. elec.	Biv. gr. hp	Biv. o. air hp
No. build. retr.	1.48	1.54	1.69	1.45	1.57	1.23	1.45	1.69	1.38	1.48
Savings:										
Attic fl. ins.	.00	.00	.01	.00	.00	.00	.00	.01	.00	.00
Floor ins.	.00	.00	.00	.00	.00	.00	.00	.00	.00	.00
Ext. wall ins.	.05	.05	.11	.04	.06	.00	.04	.11	.00	.03
Ins. wall ins.	.00	.00	.00	.00	.00	.00	.00	.00	.00	.00
Triple-glazing	.06	.07	.09	.06	.08	.04	.06	.08	.05	.06
Triple-gl. l.e.	.00	.00	.00	.00	.00	.00	.00	.00	.00	.00
Tr.-gl. l.e.g.	.00	.00	.00	.00	.00	.00	.00	.00	.00	.00
Weatherstrip	.01	.01	.02	.01	.01	.00	.01	.01	.00	.00
Exh. air h.P.	.00	.00	.00	.00	.00	.00	.00	.00	.00	.00
Sum. of retro.	1·36	1·41	1.46	1·34	1·42	1·20	1·34	1·48	1·33	1·39
Sum. of comb.	1·36	1·41	1·46	1·34	1·42	1·20	1·34	1·46	1·33	1·39
Distribution:										
Sal.'old boiler	.00	.02	.02	.02	.02	.02	.02	.02	.02	.02
New boil. cost	.08	.10	.03	.06	.28	.09	.06	.03	.25	.31
Piping cost	.00	.01	.00	.01	.16	.01	.01	.00	.07	.01
Energy cost	.60	.59	.62	.56	.28	.63	.56	.61	.34	.35
Connection fee	.00	.00	.00	.01	.00	.01	.01	.00	.00	.00
Buil. retrof. c	.43	.43	.54	.43	.43	.19	.43	.54	.40	.44
Inevitable cost	.25	.25	.25	.25	.25	.25	.25	.25	.25	.25

increased, and therefore this is not a good strategy. District heating, a ground water coupled heat pump, and a bivalent heat pump–oil boiler system are other heating systems with a lower LCC, but the best solution was natural gas. The only building retrofit to be implemented was triple-glazed windows because the old ones were dilapidated! It was also shown that the combination of the retrofit LCC and the incremental LCC has the same value for all the heating systems, with the exception of electrical heating with a time-of-use rate, which is of no interest to the optimal solution. More details and a thorough presentation of the input values for this LCC optimization are presented in Gustafsson (1990). Experience shows that it is usually optimal to use a heating system with a very low operating cost. The acquisition cost for the system, however, cannot be too high, as is the case for a heat pump only meeting the total demand in the house (Table 3.1). Note that there are only a few building and ventilation retrofits which are optimal to install and of those that are optimal the cost of them is low or otherwise their remaining life are very short.

3.8 Sensitivity analysis

In the case shown in Table 3.1, there is one solution showing a LCC much lower than the others. This is not always the situation and usually two, or more, of the strategies may be very close to each other, making it hard to

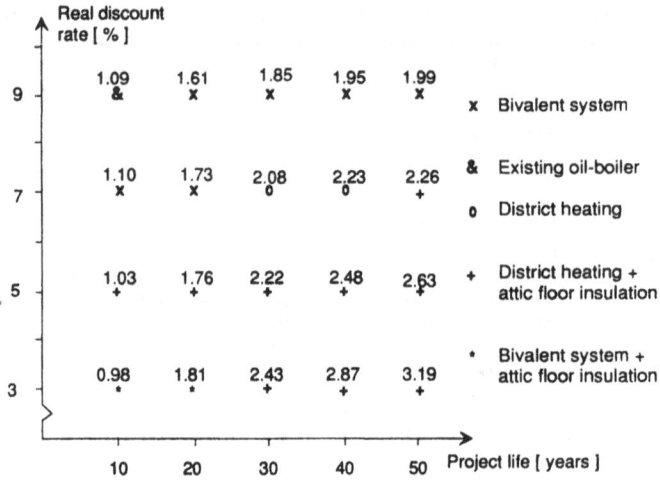

Figure 3.3   Bivariate sensitivity analysis (Gustafsson and Karlsson, 1989a).

know which one to choose. A sensitivity analysis might solve this problem. The aim with such an investigation is to determine if the optimal solution will change substantially with minor modifications of the input data. Of special interest are changes in the discount rate and the project life of the building, as these values cannot be set with total accuracy. Further, variations in energy prices must often be examined, as well as many other items in the input data file. The result may be presented by use of a bivariate diagram (Flanagan *et al.*, 1987). One example is shown in Figure 3.3, presented in Gustafsson and Karlsson (1989a). Note that the two cases in Table 3.1 and Figure 3.3 are not identical.

From Figure 3.3 it is clear that both the project life and the discount rate have a significant importance for the optimal strategy. Note also that the value of the LCC will change for different input values of the rate and project life, but this does not imply that a 3% rate and a 10-year project life are the best values to choose just because this alternative has the lowest LCC. Different strategies must be compared using the same rate, etc. It is important to note that for higher discount rates, less complicated heating systems are chosen, even if they have higher operating costs. For a 3% rate, the bivalent system, which has a very low operating cost but a high acquisition cost, is the best, while an oil boiler is optimal for a rate of 9%. Insulation measures will have an advantage in a long project life, but will be less profitable for a high discount rate. Of mostly academic interest is the fact that the LCC will almost always get lower for higher discount rates, but this fact is not valid for very short project lives. For a project life of 10 years, the LCC is increased when the rate is increased from 3 to 5%. This fact is dealt with in more detail in Gustafsson (1988). In Sweden, district heating is provided by burning a mix of fuels in the utility plant. During the summer, most of the heat comes from burning refuse in an incineration plant, while oil or coal must be used in the winter. The cost for district heating is consequently lower than the oil price, while, at the same time, the installation cost is higher than the cost of an oil boiler. This is why it is optimal to use district heating for some combinations of discount rates and project lives. It must also be noted that the amount of additional attic insulation is not the same in the optimal strategies shown in Figure 3.3. Longer project lives and lower discount rates imply more insulation. Also, the optimal thickness of insulation is not a continuous function. When it is optimal to add insulation it is often necessary to apply more than 0.1 m or it may be better to leave the building as it is (Gustafsson and Karlsson, 1990). The same reference also emphasizes the importance of the remaining life of the building asset. If this is very short it will often be optimal to add extra insulation, and in that case an extensive amount of insulation should be chosen, say 0.2 m. Such a measure will very much decrease the heat flow through (for example) a wall, and this will imply that if all retrofits are made when they are unavoidable, the thermal state of the building will become better and better, and the cost for achieving this will be lower than leaving the building unchanged.

The influence on input data changes may be split into three different categories: (a) where the LCC will increase for an increase in input data; (b) where the LCC will decrease for an increase; and (c) where the LCC will not change at all for changes in the input data. Some examples of the first category are changes in building costs, installation costs, etc. To the second category apply changes in, for example, the discount rate, the remaining life of a building asset and the outdoor temperature. Some of the input data will apply to more than one of the categories. Consider, for example, a small increase in the cost of the oil boiler. If the oil boiler is part of the optimal solution, the LCC will increase if the cost for the boiler is increased. However, when the cost passes a certain limit the oil boiler will fall out of the solution, and from that point further increases in the oil boiler cost are of no interest. This fact is often used in the practical work with life cycle costing. When a building is analysed for the first time, input values can be chosen without a tedious examination process. The important thing is that the chosen values, at least to some degree, will reflect the real situation. After the first optimization has been elaborated, only the strategies that are close to each other need to be scrutinized. This means that much of the initial work with the input data might not be necessary and that only some of the details must be examined more closely.

In Gustafsson (1988), a sensitivity analysis of all the values used in an optimization is elaborated, but it is not possible to repeat this here. Some of the facts found must, however, be mentioned. For instance, it could be assumed that a small change in the resulting LCC will not be as important as if larger changes are encountered. This is not always true. If a 5% change in the discount rate was introduced, this could lead to about a 2% change in the LCC, which is one of the largest differences found. However, the LCC for the existing building also changes by approximately the same amount, and implies that the optimal strategy will be almost the same for small changes in the discount rate.

A very high existing U-value for, say, an external wall in a poor thermal status might be expected to influence the LCC very much and to influence the new optimal U-value. This is not so. The new optimal U-value is not influenced by the existing one (Bagatin *et al.*, 1984; Gustafsson 1988), and the fact is that as long as the optimal insulation is introduced, the resulting LCC is almost constant. The same is valid for the actual insulation cost. If this cost is increased the optimization results in a thinner insulation which in turn will decrease the new LCC.

Annual increases in energy prices will naturally lead to a more extensive retrofit strategy which will lead to a lower LCC than might first be expected. This will also imply that, if the proprietor knows in advance what the energy prices will become, there is a better possibility of reducing the effects of escalating energy prices than if no action is taken at all. In some way, the optimization leads to a model that is in some regards self-regulating. The optimization makes the best of the situation and the result of a change might not be as bad as first assumed.

3.9   Linear programming techniques

In recent years there has been an increased interest in linear programming. The technique, which was developed in the 1960s, is not in common practice, because of the very tedious calculation procedures, and the use of fairly advanced mathematics. However, now that computers are on every desk, the situation is different, and the design of mathematical software makes the solving of complex linear programs much easier. It must be noticed that linear programming is an optimization technique which is not confined to life cycle costing. The reason for choosing linear programming is that it is possible to prove mathematically that the optimum solution, i.e. the best solution with the lowest LCC, has been found. The method is also suitable when discrete time or cost steps are included in the problem. This might seem to be of only minor interest but the tariffs for energy for tomorrow will probably always be of the time-of-use type where the price changes according to the time of day or year. In traditional methods, such as OPERA, these tariffs must be normalized many times and approximated by a mean value of the real price, which might greatly influence the optimal solution (Gustafsson and Karlsson, 1988). It is not possible to deal with linear programming in detail here and thus only a very brief presentation is made.

The LCC must be expressed in a so-called objective function. This function, which is the expression to be minimized, must be totally linear, i.e. it is not possible to multiply or divide two variables by each other. A variable must only be multiplied by a constant. The objective function is, after this, minimized under a set of constraints which also have to be linear functions. All of the constraints must be valid at the same time. The procedure for solving such problems includes the use of vector algebra and is not dealt with here. (See Foulds (1981) for the basic concepts, and Murtagh (1981) for deeper insights into linear programming and how to solve such problems.)

In Sweden, it is common to describe the climatic conditions for a site by the use of mean values of the outdoor temperatures for each month of the year. Using twelve mean values instead of a continuous function makes it possible to use the linear programming technique, as it is not possible to take derivatives of functions with discrete steps. The thermal load in kilowatts and the need for heat in kilowatt-hours will consequently also follow the climate function, which implies that the steps are also included when the thermal situation is elaborated. In Table 3.2 the initial thermal load is shown for a building in Malmö, Sweden.

Suppose that only the attic floor insulation is of interest, in order to make the problem shorter and easier to deal with. The new demand for the building is now to be calculated. One variable is thus introduced showing the thermal load, in kilowatts, for the building for each month. Further, suppose that the building is heated by district heating where a time-of-use tariff is used. The cost for heat is assumed to be 0.2 SEK/kWh during November to March and 0.10 SEK/kWh for the remaining months. The first part of an objective function might be presented as:

$$(H_1 \cdot 744 \cdot 0.2 + H_2 \cdot 678 \cdot 0.2 + H_3 \cdot 744 \cdot 0.2 + H_4 \cdot 720 \cdot 0.1 \qquad (3.6)$$
$$+ \ldots + H_{12} \cdot 744 \cdot 0.2) \cdot 18.26$$

where H = the new optimal heat load in kW for each month, $1, 2, \ldots$ = the number of the month, $744, \ldots$ = the number of hours in each month, $0.2, 0.1$ = the district heat price for various months in SEK/kWh and 18.26 = the present value factor.

Note that the influence of leap years is considered for February. The demand in Table 3.2 must be covered in one way or another. The model is

Table 3.2 Heat demand for a building sited in Malmö, Sweden

Month	Heat (MWh)	Month	Heat (MWh)	Month	Heat (MWh)
January	32.60	May	15.95	September	12.02
February	30.95	June	9.92	October	18.99
March	29.85	July	6.97	November	24.07
April	22.53	August	7.70	December	28.98

therefore supplemented by twelve constraints showing the situation for each month and the first three of them will be:

$$H_1 \cdot 744 \geqslant 32.60 \qquad H_2 \cdot 678 \geqslant 30.95 \qquad H_3 \cdot 744 \geqslant 29.85 \qquad (3.7)$$

The cost for additional insulation was shown in equation (3.3) and the influence that this insulation has on the thermal load in equation (3.4). From equation (3.4) it is obvious that it is not a linear expression, since t_{ins} is present in the denominator. However, it is possible to make it a linear function of t_{ins}, but in that case equation (3.3) will be nonlinear. A method described by Foulds (1981), called piecewise linearization, is therefore used. In this method, the value of a function is calculated for a number of discrete values of t_{ins} and each value for the function is coupled with a binary integer variable which can only have the value 1 or 0. All these binary variables are added and the sum is constrained as lower than or equal to 1. This forces the model to choose one or none of the variables. The original nonlinear function of t_{ins} is thus changed to a linear function of the binary variables. The situation is depicted by the following example. The decrease of the heat demand is shown in equation (3.4), and for five steps of insulation thickness, the decrease is as presented in Table 3.3. (See also Gustafsson and Karlsson (1989b).)

Suppose the area of the attic floor is $200 \, m^2$. The number of degree hours in Malmö for January has been calculated as 15 996 and subsequently the decrease in heat flow, in kWh, through the attic will become:

$$10^{-3} \cdot 15\,996 \cdot 200 \cdot (0.4 \cdot A_1 + 0.533 \cdot A_2 + 0.6 \cdot A_3 \qquad (3.8)$$
$$+ 0.640 \cdot A_4 + 0.667 \cdot A_5)$$

Equation (3.8) and eleven more expressions for the rest of the year must be added to the left-hand sides of the constraints in equation (3.7). Note also that:

$$A_1 + A_2 + A_3 + A_4 + A_5 \leqslant 1 \qquad (3.9)$$

and that the A variables are all binary integers. One or none of them must be chosen according to equation (3.9). Only lacking now is the building cost for the additional insulation. Using the same values as above for derivative

Table 3.3 Decrease in U-value for five discrete steps of additional insulation

Added insulation (m)	Variable	Existing U-value*	New U-value*	Decrease in U-value*
0.05	A_1	0.8	0.400	0.400
0.10	A_2	0.8	0.267	0.533
0.15	A_3	0.8	0.200	0.600
0.20	A_4	0.8	0.160	0.640
0.25	A_5	0.8	0.133	0.667

* In $Wm^{-2}K^{-1}$.

optimization, the cost will be as a function of A_1-A_6 instead of t_{ins}:

$$200 \cdot [(100 + 0.05 \cdot 600) \cdot A_1 (100 + 0.10 \cdot 600) \cdot A_2 + \dots \qquad (3.10)$$
$$+ (100 + 0.25 \cdot 600) \cdot A_5]$$

The model is now totally linear and it is possible to use ordinary linear or mixed integer programming methods for optimization. By the use of more binary integers it is again possible to add the influence of the inevitable retrofit cost as well, i.e. when one of the A variables is chosen, a certain amount is added to the objective, and if none is chosen a different amount is added. As can be found from the above example, the number of equations and constraints will become very large for real-world problems. Now, the tedious work of generating equations and constraints is dealt with by separate computer programs which are used for writing the large input data files. More details and a complete model can be found in Gustafsson (1992).

3.10 Summary

Two different methods are shown for optimizing the retrofit strategy for a building.

In the first method the LCC is actually calculated for a number of cases and the lowest LCC strategy is selected. The other method shows how to design a mathematical model in the form of mixed integer programming. The latter method demands a more skilled mathematician because of the use of vector algebra when solving the problem. However, there are advantages using this latter method owing to the possibilities of solving discrete problems, i.e. the functions are not necessarily continuous. One major drawback is that the problems to be solved must be totally linear, but by the use of piecewise linearization this drawback can be dealt with, at least to some extent.

References

Bagatin, M., Caldour, R. and Gottardi, G. (1984) *International Journal of Energy Research*, **8**, 127–138.

Flanagan, R., Kendell, A., Norman, G. and Robinson, G. (1987) Life cycle costing and risk management, paper presented at the CIB–1987 Conference, Copenhagen, Denmark.

Flanagan, R., Norman, G., Meadows, J. and Robinson, G. (1989) *Life Cycle Costing, Theory and Practice*, BSP Professional Books, Oxford.

Foulds, L.R. (1981) *Optimization Techniques*, Springer–Verlag, New York.

Gustafsson, S.-I. (1986) Optimal energy retrofits on existing multi-family buildings, Thesis No. 91, LIU-TEC-LIC-1986: 31, Institute of Technology, Linköping, Sweden.

Gustafsson, S.-I. (1988) The OPERA model. Optimal energy retrofits in multi-family residences, Dissertation No. 180, Institute of Technology, Linköping, Sweden.

Gustafsson, S.-I. and Karlsson, B.G. (1989a) Life cycle cost minimization considering retrofits in multi-family residences, *Energy and Buildings*, **14**(1), 9–17.

Gustafsson, S.-I. (1990) A computer model for optimal energy retrofits in multi-family buildings.

The OPERA model, Document D21: 1990, Swedish Council for Building Research, Stockholm, Sweden.

Gustafsson, S.-I. (1992) Optimization of building retrofits in a combined heat and power Network, *Energy*, **17**(2), 161–171.

Gustafsson, S.-I. and Karlsson, B.G. (1988a) Why is life cycle costing important when retrofitting buildings? *International Journal of Energy Research*, **12**, 233–242.

Gustafsson, S.-I. and Karlsson, B.G. (1988b) Bivalent heating systems, retrofits and minimized life cycle costs for multi-family residences, in: *Proc. CIB W67 meeting*, Bulletin No. 153, The Royal Institute of Technology, Div. of Building Technology, Stockholm, Sweden, pp. 63–74.

Gustafsson, S.-I. and Karlsson, B.G. (1989b) Insulation and bivalent heating system optimization: Residential housing retrofits and time-of-use tariffs for electricity, *Applied Energy*, **34**, 303–315.

Gustafsson, S.-I. and Karlsson, B.G. (1990) Energy conservation and optimal retrofits in multi-family buildings. *Energy Systems and Policy*, Vol. 14, pp. 37–49.

Gustafsson, S.-I. and Karlsson, B.G. (1991) Window retrofits and life cycle costing, *Applied Energy*, **39**(1), 21–29.

Markus, T.A. (1979) The window as an element in the building envelope; techniques for optimization, in: *Proc. 1979 CIB Conference*, vol. 2, Copenhagen, Denmark, pp. 255–268.

Marshall, H. (1989) Review of economic methods and risk analysis techniques for evaluating building investments, *Building Research and Practice*, **17**(6), 343–349.

Murtagh, B.A. (1981) *Advanced Linear Programming: Computation and Practice*, McGraw-Hill.

Ruegg, R.T. and Petersen, S.R. (1987) *Least-Cost Energy Decisions*, NBS Special Publication No. 709, National Bureau of Standards, Washington, DC, USA.

Sonderegger, R., Cleary, P., Garnier, J. and Dixon, J. (1983) *CIRA Economic Optimization Methodology*, Lawrence Berkeley Laboratory, Berkeley, CA, USA.

Van Dyke, J. and Hu, P. (1989) Determinants of variation in calculating a discount rate, *Energy International Journal*, **14**(10), 661–666.

4 Life cycle costing of highways
R. ROBINSON

4.1 Introduction

4.1.1 Choosing between highway investments

Investing significant sums of money in building new roads, or improving those that already exist, requires careful appraisal to ensure that optimum use is being made of the sums invested. Investment choices in the highways subsector involve making decisions about route choice, geotechnics, earthworks, geometric and pavement design, drainage and the design of structures.

Making these choices on a rational basis requires the comparison of different levels of investment at the present time compared with their respective consequential future costs. The adoption of higher design standards normally leads to higher initial costs, but may result in lower costs to the highway agency in terms of future costs of maintenance and renewal. Such considerations of life cycle costs take on another dimension when considering strategic issues on national roads, when it also becomes appropriate to consider ongoing costs to roads users. These may consist of vehicle operating costs, time and delay costs, and the costs of road accidents.

If life cycle costs are not taken into account, investment decisions become subjective and dependent on the application of standards that are often themselves based on historical precedent rather than objective analysis. The rational formulation of standards should also depend on life cycle cost considerations.

Costs that must be taken into account when considering roads over their life cycle fall into two main groups: those affecting the highway agency, and those affecting road users.

4.1.2 Highway agency costs

Costs incurred by highway agencies include the ongoing disbursements for maintenance in all its forms to:

- pavements
- footways and footpaths
- cycletracks
- drainage features

- structures
- signs and signalling.

Some of these are fixed costs, being dependent only on the policy of the agency responsible for maintenance; others depend on the environment in which the road is situated; others depend on the volume, intensity and loading of traffic using the road; others are time-related.

For all new roads, or for major reconstructions or upgradings, costs of construction need to be considered. These costs include those for planning, design, procurement, the construction itself, and its supervision and management. They will depend on several factors:

- the standard of road being constructed, from motorway to housing-estate road
- the geographical location within the country
- the geotechnical environment through which the road will pass, including topography, soils, etc.
- the degree of urbanisation surrounding the road—this will affect, in particular, the need for structures
- the 'sensitivity' of the natural and built environment through which the road will pass—this will influence the extent of measures necessary to mitigate any environmental damage.

There is often a trade-off between construction and maintenance costs. Clearly, the higher the design standard, the stronger will be the road pavement and the more resistant it will be to deterioration. A principal use of life cycle costing is to investigate standards and policies that optimise the balance of resources between these two aspects of investment.

The determination of construction and maintenance costs is discussed later.

4.1.3   Road user costs

Costs to the road user are normally considered to include: .

- vehicle operating costs, including both running and standing costs
- time costs, including those for delays due to congestion and roadworks
- road accident costs.

Of these, vehicle operating costs are relatively easy to determine but, although amounts of time and numbers of accidents associated with a particular road may be determined relatively easily, their valuation is often contentious. These issues will be discussed further.

In the same way that a life cycle cost approach may be used to investigate trade-offs between road construction and maintenance costs, the approach can also take road user costs into account to assist in determining optimum standards of design and policies of road maintenance. Indeed, when carrying out an economic analysis of a project, a viewpoint is normally taken within the context of the national (or even international) economy. In such cases, the inclusion of road user costs in the analysis is an essential ingredient, since funding is normally provided from national and local taxation or duties.

Over the life of a road, road user costs generally far exceed the costs of construction and maintenance incurred by the highway agency. For typical roads in developing countries, road user costs can represent up to about 80% of the total transport costs over the project life; construction costs are of the order of 20% of total costs, whereas road maintenance costs account for only a few per cent. Figures in industrialised countries are likely to be similar, although there will be considerable variation between roads of different classes and hierarchy.

These costs are not independent of one another. Vehicle operating costs depend on the number and type of vehicles using the road, the type of journey that they make, the geometry of the alignment, and the condition of the road surface. For instance, if the alignment has many steep gradients, fuel consumption will obviously be higher than for a flat road. The geometry also affects directly the cost of construction. In hilly terrain, the cost of building wide roads with flat gradients is high. Construction costs can be reduced by building a narrower road with steeper gradients, but this will be at the expense of a higher cost of operating vehicles.

Time costs will also be related to road geometry, since a road built to a high standard will allow higher travel speeds. Similarly, restricted-access highways will reduce delays at junctions and in urban areas. Road accident rates are also related to the number of junctions and the skid resistance of the road surface. Both of these features are a function of design standard and will affect construction cost. Surface characteristics will affect vehicle speeds and operating costs, and these will also be a function of road maintenance policy and cost.

4.1.4   Other costs

Other costs can also be taken into account when analysing road schemes on a life cycle basis. These may include environmental costs, institutional costs and other consequential costs associated with the provision of the road. It may also be appropriate to include some benefits in a full cost analysis, such as agricultural, industrial or developmental benefits. These are often difficult to quantify, and will not be discussed further, except those which manifest themselves as generated traffic.

4.1.5   Changing costs over time

The analysis of life cycle costs for roads assumes increased complexity because costs and the relationship between costs do not stay constant over time. As time passes, roads deteriorate. As levels of deterioration increase, the need for road maintenance, and hence its cost, increases. However, with deterioration, roads become rougher, and vehicle components, such as tyres and mechanical parts, will wear out more quickly, resulting in higher operating costs. Likewise, very rough roads, such as are often found in developing countries, may force vehicles to reduce speeds, or increase the risk of accidents, leading to higher costs in both of these cases.

Thus, all road user costs are affected by the condition of the road surface and will change with time as it deteriorates, is maintained or is rehabilitated. The condition of the road surface, besides being affected by the design and construction standard, is also affected by the traffic loading, the standards (and hence the costs) of maintenance, and the environment. The more vehicles that use the road and the heavier their axle loads, the more quickly will the road wear out. The road will also wear out more quickly if it is subject to extremes of climate, which may weaken the structure or cause erosion.

However, the rate of deterioration of the road's structure and its surface condition can be slowed down by effective and timely maintenance at intervals throughout the road's life. Thus, environment, traffic and maintenance all affect the surface condition of the road and therefore all have an effect on the cost and changes in cost experienced by the road users.

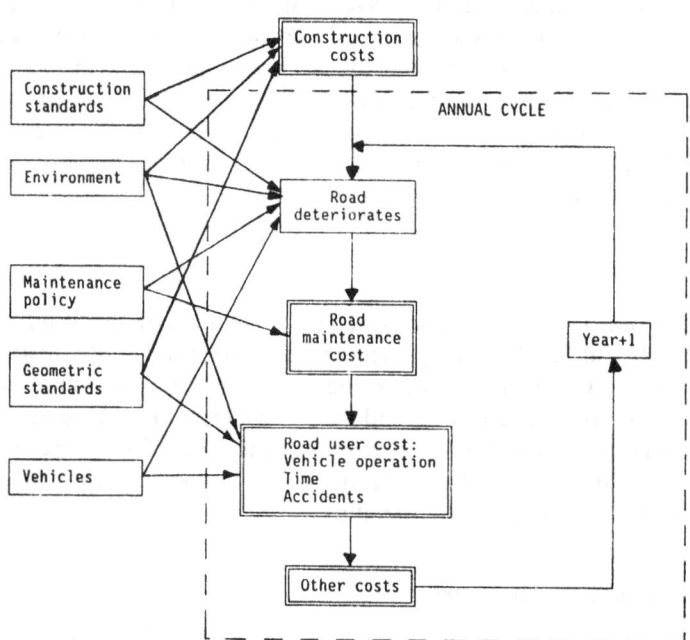

Figure 4.1   Annual cycle of cost and road deterioration.

It is convenient to consider this cost and deterioration cycle in the schematic way shown in Figure 4.1. It is this approach that is fundamental to the formation of most life cycle cost models used in the roads subsector.

4.1.6 The relevance of life cycle costing to highways

In the highways subsector, recurrent vehicle operating costs are high compared with the costs of initial construction. A life cycle costing approach to decision making is therefore considered essential if the quality of decisions is not to be biased by consideration only of short-term issues leading to detrimental longer-term consequences and avoidable higher costs. This contrasts with the recommendations by Ferry and Flanagan (1991) in a CIRIA report. They cast doubts on the use of a life cycle cost approach except for those assets which have a relatively short life, high recurrent costs, and a formalised mandatory maintenance programme. Their conclusions appear to be based largely on experience in the water industry and they make scant reference to the large body of material and experience relating to highways. As noted above, highways comply with at least two of Ferry and Flanagan's criteria and, therefore, their conclusion is considered not relevant to highways.

4.2 Historical background

4.2.1 City of London, 1870

It is possible to trace the use of life cycle costing of highways back well before the 1890s. Croney (1977) reports that, prior to 1870, engineers in the City of London used records extending back over 40 years to determine the life cycle costs of stone sett road pavements for comparison with the more commonly used water-bound macadam construction. The whole life costs were evaluated for a variety of traffic conditions by combining both initial construction costs and annual maintenance costs using a discounted cash-flow technique. Stone sett paving proved to have a higher initial cost but low maintenance cost under heavy traffic, so, as a result, it was used in locations such as the approaches to the London docks. On less heavily trafficked roads, water-bound macadam construction was adopted on the basis of its lower life cycle costs under these conditions.

4.2.2 Road Reasearch Laboratory, 1969

A report was produced by the then Road Research Laboratory (1969) which compared the costs of constructing and maintaining flexible (bituminous) and rigid (concrete) pavements over 50 years. Estimates were made of initial construction costs, and subsequent maintenance and user delay costs for four types of road: a rural motorway, a peri-urban road, a rural secondary road and a road on a housing estate. Costs were discounted to present-day values for various forms of rigid and flexible road pavements. The report was used to develop a method of assessing some of the life cycle costs of different forms of construction so that these factors could be taken into account when making decisions on the award of contracts.

4.2.3 The MIT model

Major advances in life cycle costing for roads were instigated by the World Bank, which was seeking methods of improving the quality of investment decisions in this subsector in the late 1960s. The first step was to produce draft terms of reference for a 'highway design study' for internal consideration within the Bank. This was then followed by commissioning a group at the Massachusetts Institute of Technology (MIT) to carry out a literature survey and to construct a computer model based on information already available. The highway cost model produced by MIT (Moavenzadeh *et al.*, 1971; Moavenzadeh, 1972), was a considerable advance over any other methods existing at the time for examining the interactions between road construction costs, maintenance costs, and vehicle operating costs. However, the model highlighted areas where more research was needed to replace relationships that were inappropriate to developing-country environments, and to provide additional relationships.

4.2.4 The TRRL road investment model (RTIM)

In order to address deficiencies in data highlighted by the MIT model, the then Transport and Road Research Laboratory (TRRL), in collaboration with the World Bank, undertook a major field study in Kenya to investigate the deterioration of paved and unpaved roads, and the factors affecting vehicle operating costs in a developing country. The performance of more than ninety 1-km long test sections of road was monitored at regular intervals over a period of two years. An experiment was carried out in parallel to measure vehicle speeds and fuel consumptions over the same test sections. In addition, data were collected from many commercial vehicle operators on such items as spare parts, usage of vehicles, maintenance and labour requirements, tyre wear and vehicle depreciation. Relationships were then developed relating these directly to vehicle operating conditions (Abaynayaka *et al.* 1977).

The results of this study were used to calibrate a prototype computer model called RTIM (Robinson *et al.*, 1975) which could be used for evaluating the life cycle costs of construction, maintenance costs and vehicle operating costs for a road investment project in developing countries. This prototype was tested extensively and proved to represent a significant advance in road planning methodology in developing countries, based on life cycle cost principles.

Experience gained by TRRL in use of the prototype RTIM suggested that users in developing countries had a need for a simpler approach than that offered by this mainframe-based computer program. Some of these difficulties were overcome by publishing the relationships built into the model as a book of tables (Abaynayaka et al., 1976). This enabled users in developing countries, who did not have access to computers, to utilise the relationships in the model. In addition, the RTIM model was reprogrammed to make it easier to use and to fit on to a smaller computer (Parsley and Robinson, 1982). In reprogramming the model, the opportunity was also taken to incorporate the results of the latest research into vehicle operating costs carried out by TRRL in the Caribbean (Morosiuk and Abaynayaka, 1982; Hide, 1982). With the subsequent advent of microcomputers, a version of RTIM was later developed to run on such machines.

4.2.5   The World Bank's HDM model

Following the original work on RTIM, the World Bank took a different view from TRRL about the needs for further development. Rather than adopting a simpler approach that was easier to use, they saw the need for a more complex model that could carry out analysis of multiple road links in parallel, would have a more powerful economic analysis capability, and could undertake automatic senstivity analysis of key variables such as discount rate and traffic growth. In 1976 a contract was awarded to MIT to produce the highway design and maintenance standards model, HDM (Harral et al., 1979).

The HDM proved to be a powerful analytical tool, but drew attention to the limitations of the relationships for road deterioration and vehicle operating costs which were based on TRRL's pioneering work in Kenya. As a result of this, the Bank embarked upon a $19 million research programme in conjunction with the Brazilian government (Harrison and Chesher, 1983), and supported a similar study being carried out in India by the Central Road Research Institute (CRRI, 1982) to try to improve the quality of the relationships by extending them to a wider range of vehicles and road construction types.

The results of all these studies were incorporated into the third version of the Bank's computerised highway subsector planning and investment model (HDM-III), and a five-volume series of reports was produced which represented the culmination of an 18-year endeavour in this area of life cycle costing (Chesher and Harrison, 1987; Paterson, 1987; Watanatada et al., 1987a, b).

4.2.6   The Department of Transport's COBA and URECA models

Life cycle costing was introduced by the Department of Transport (DoT) of the UK for the appraisal of major road investments in the 1970s with the development of their cost benefit analysis (COBA) computer program. COBA compares the costs of road schemes with the benefits which can be derived by road users, and expresses the results in terms of a monetary valuation. The approach recognises that monetary values cannot be put on all costs and benefits, which, for a road scheme in an industrialised country, may include factors such as visual intrusion or changes in the number of people affected by traffic noise. For major road investments, the DTp considers the non-monetary terms within a 'framework', where the differences between the effects of different options can be compared.

COBA is limited to those factors that are relatively easy to value, which confines its coverage of the benefits to those gained by road users. Benefits are compared for the road system with or without the scheme, with user costs including:

- journey times on links and at junctions
- accidents
- vehicle operating costs.

On the expenditure side, COBA takes account of those elements which fall directly on the financing authority. These costs are:

- capital costs, including construction, land, preparation and supervision costs
- maintenance costs.

COBA evaluates costs and benefits over a 30-year period and bases results on the net present value test. It is used routinely in the United Kingdom for the analysis of all major interurban road schemes submitted to the DTp for funding. The current version of the program is COBA9 (Department of Transport, 1981). Recently, the Department has introduced a new program, URECA, for the assessment of urban roads (Stark, 1990).

4.2.7   The TRRL whole life cost model

Concerns about the lack of a life cycle costing approach to road maintenance in the United Kingdom in the mid 1980s (Garrett, 1985) led the then TRRL to undertake the development of a whole life cost model (WLCM) using relationships based on full-scale field trials and the monitoring of highway networks that had been undertaken by the laboratory over many years. Originally time-dependent and traffic-dependent versions were available for flexible carriageways, and a separate version was available for rigid carriageways (Abell et al., 1986; Kilbourn and Abell, 1988). More recently, these have been combined into one model (Abell, 1992).

The model is applicable to roads in the United Kingdom and differs from HDM-III in terms of the pavement construction types included, and its calibration to pavements subject to freeze–thaw cycles rather than those in a tropical climate. It uses 85th percentile design curves for all designs rather than the mean value approach to forecasting in the World Bank's model. Differences between WLCM and HDM-III have been evaluated by Petts and Brooks (1986).

4.2.8   Other approaches to life cycle costing

Several other models have been developed around the world to assist in the evaluation of life cycle costs for highways. These include: EAROMAR (Markow, 1984), LIFE2 (Lindow, 1978), NIMPAC (Both and Bayley, 1976), RENU (Garcia-Diaz et al., 1981), to name but a few. Other approaches to life cycle costing for roads have been described in the technical literature, including those by Barber et al. (1978), Feighan et al. (1981), Findakly et al. (1974), Jung (1985), Litten and Johnston (1979), Loong (1989), Ockwell (1990), Rada et al. (1986), and Sullivan and Scott (1990).

4.3   Traffic

Traffic demand is the key variable on which the life cycle cost analysis of highways depends. Variations in traffic over the road's life affect the economic justification of road improvements, the annual amounts to be spent on maintaining each section of the road, the selection of road design standards and the magnitude of construction costs (Howe, 1973). Determining values for traffic parameters requires the estimation of baseline traffic flows and the forecasting of flows into the future.

Traffic is normally categorised, for life cycle costing purposes, in terms of:

- normal
- diverted
- generated.

Normal traffic is that which would use the road even if no improvements were undertaken and is usually defined in terms of the annual average daily traffic (AADT). Estimates are commonly based on the results of traffic counts using automatic traffic counters. These are often supplemented by manual counts in order to determine the breakdown of flow by vehicle classification.

Diverted traffic is that which diverts from another route, network or mode as a result of the improvement undertaken. Estimates of diverted traffic should be based on the results of origin and destination surveys. The simplest method of estimating diversion is to assume that all vehicles that would save time or money by diverting would do so.

Generated traffic is that which chooses to make a journey because lower costs result from the provision or the improvement of the facility. These lower costs are usually due to reduced journey times or vehicle operating-cost savings. Forecasting generated traffic can be very difficult, but is normally done using demand curves, such as that shown in Figure 4.2.

As cost is reduced from C_1 to C_2, traffic flows increase from T_1 to T_2. The amount of this increase is determined by the shape of the 'demand curve'. The slope of this is termed the 'elasticity of demand' (e). Results from unpublished research by the World Bank and by White (1984) suggest that, world-wide, the price elasticity of demand for passenger transport varies between -0.6 and -2.0, with an average of about -1.0. The elasticity of demand for goods transport has been found to be very much lower, and depends on the proportion of transport cost in the commodity price. Because of the difficulty of determining the demand curve, it is often replaced by a straight line with slope e. In most cases, this simplification does not lead to a significant increase in the error, particularly given the general uncertainty of the estimating process.

For life cycle costing purposes, it is necessary to forecast traffic levels into the future. The breakdown of flow into normal, diverted and generated traffic assists with this. Diversion and generation are, usually, once-and-for-all effects and, after the first year, these types of traffic can be assumed to grow at the same rate as normal traffic. Forecasting future traffic levels is an uncertain process and compounds errors obtained when estimating annual flows based on short-duration sample counts.

Traffic flows have cycles which vary on an intra-day, intra-week and intra-season basis. Unless counts are carried out for an entire year, forecasts of AADT must be based on sample counts over shorter durations. It has been

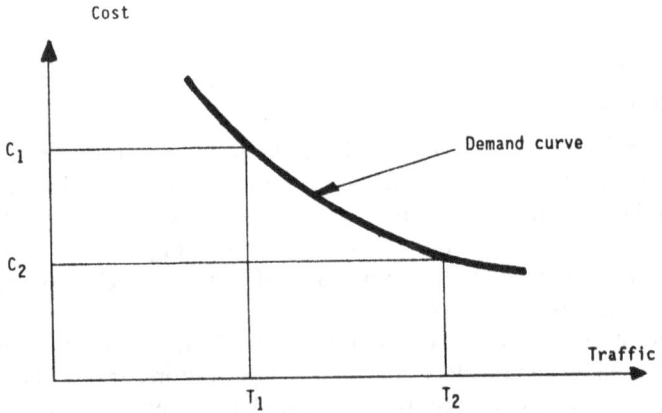

Figure 4.2   Use of demand curve for predicting generated traffic.

shown by Howe (1972), for example, that such estimates can be subject to large errors, particularly where flows are very low. Given this initial uncertainty, it is not surprising that forecasting future flows is an even more uncertain process. This is the case in industrialised countries with developed and stable economies, but in developing countries the problem becomes more intractable. The economies of these countries are often very sensitive to world prices of just one or two particular commodities, and fluctuations in world oil prices and supply in recent times have added a new dimension to the difficulties.

Forecasts of future flows can be based on forecasts of gross domestic product (GDP), which take into account explicit changes in overall economic activity, or by the linear extrapolation of past trends. Whatever method is used for the forecast, it is normal to consider life cycle costs determined using 'optimistic' and 'pessimistic' traffic growth rates which span the likely range of values.

4.4 Construction costs

Of the construction costs noted earlier, those for planning, design, procurement, supervision and management are sensibly independent of life cycle cost considerations. However, it has been noted that policies and decisions on design and construction standards will have a knock-on effect on road maintenance and user costs over the life of the road. Those aspects of construction that will have most impact are:

- road geometry (earthworks and structures)
- pavement design
- drainage and structures.

4.4.1 Road geometry

Comprehensive models exist for calculating earthworks quantities and are used by designers in the preparation of tender documents and by contractors in the preparation of their bids. These are generally not appropriate for use in life cycle cost analysis because their data requirements are high and their level of accuracy is out of proportion to that which is achievable on the user cost side of the analysis. More typical of the examples of the costing methods available are those used in the World Bank's HDM-III model, which are very simple, and that used in TRRL's RTIM, which is more detailed and accurate.

The HDM approach (Watanatada *et al.*, 1987*a*) predicts earthworks costs as a function of terrain severity and geometric standards, using relationships developed as part of a Master of Science degree dissertation at MIT by W.B. Aw. For new construction, the following relationship is used to estimate

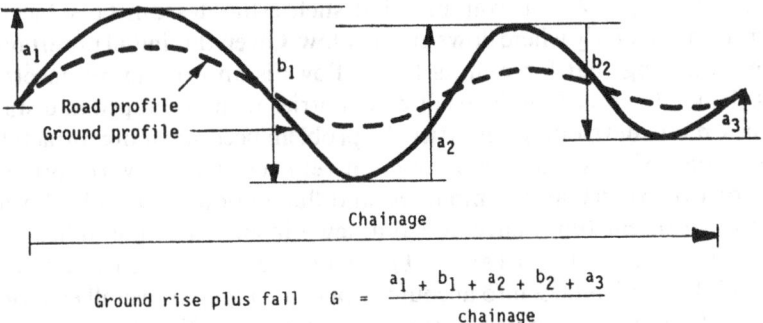

$$G = \frac{a_1 + b_1 + a_2 + b_2 + a_3}{\text{chainage}}$$

Ground rise plus fall

Figure 4.3 Determination of ground rise plus fall (Watanatada *et al.*, 1878a).

earthworks volumes:

$$E = 1000(W + 0.731H)H \qquad (4.1)$$

where E = the volume of earthworks per unit length of road, in m^3/km (including cut, fill, borrow and spoil), W = the road width, in metres, and H = the effective height of earthworks, in metres, given by:

$$H = 1.41 + 0.129(G - R) + 0.0139G$$

where G = the ground rise plus fall, in m/km (see Figure 4.3) and R = the road rise plus fall, in m/km.

The road rise plus fall is defined in the same way as that for the ground, except that the vertical profile of the road is used instead.

For road-widening projects, the earthworks volume is given by:

$$E = 1000 H \cdot W^1 \qquad (4.2)$$

where W^1 = the increase in road width, in metres.

The approach used in the RTIM model (Parsley and Robinson, 1982) recommends that a suitable vertical alignment is generated automatically from details of the ground longitudinal section using program VENUS (Robinson, 1976a). This method uses a heuristic method of ground smoothing to produce vertical alignments in conventional form and can operate with data extracted from contour maps. RTIM then uses the 'end area' method for calculating earthworks volumes and an innovative method of determining mass-haul quantities described by Davies (1972).

The RTIM earthworks model is particularly appropriate to rolling terrain where balancing earthworks is normally an objective. The method determines road centreline and ground levels at intervals along the length of road, and determines cross-sections of the form shown in Figure 4.4. Volumes of cut and fill between each pair of cross-sections are calculated by multiplying the average of the end areas of each cross-section by the distance in between them. The RTIM method can handle cases of cut–fill transitions either within

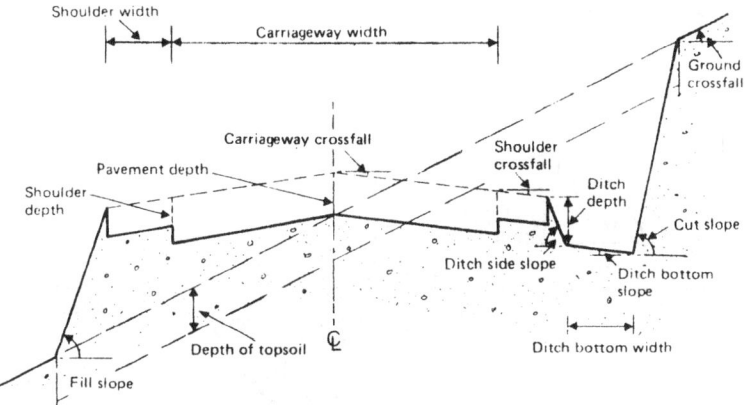

Figure 4.4 Road cross-section (Parsley and Robinson, 1982).

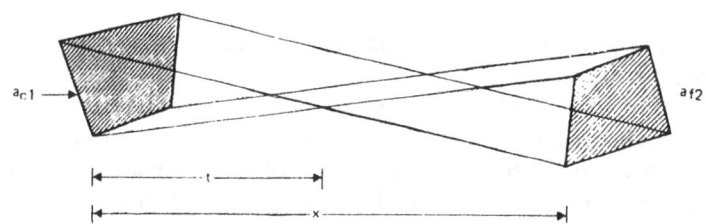

Figure 4.5 Longitudinal cut–fill transition (Parsley and Robinson, 1982).

a cross-section, as in Figure 4.4, or between cross-sections, as shown in Figure 4.5. Retaining walls can also be accommodated.

Haulage costs are calculated by using a mass-haul diagram and are based on the fact that it is never economical to transport material further than the 'marginal haul'. This is the distance at which it becomes cheaper to spoil the cut material and to borrow the fill, and is calculated from the unit cost of these operations:

$$\text{Marginal haul} = \frac{\text{unit borrow cost} + \text{unit spoil cost}}{\text{unit haulage cost}} \quad (4.3)$$

A mass-haul diagram is set up by plotting the accumulated earthworks balance (cut minus fill) against chainage, as in Figure 4.6. At each cross-section, the volume of suitable cut material is found by reducing the total cut volume by the percentage of material which is unsuitable for use as fill. Similarly, the fill volume is divided by the product of the bulking and compaction factors to give the effective volume of fill material that will be required to fill the voids. The earthworks balance for this cross-section is then the volume of suitable cut minus the effective volume of fill. To obtain the accumulated balance, this figure is added to the accumulated earthworks

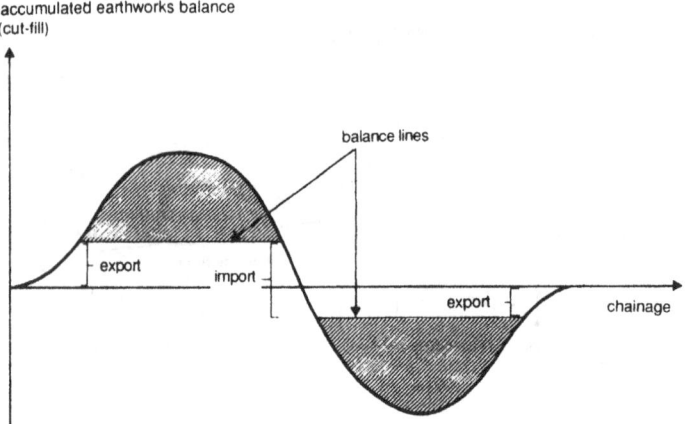

Figure 4.6   Mass-haul diagram (Parsley and Robinson, 1982).

balance for the previous cross-section. This is plotted against chainage to give the mass-haul diagram.

The procedure in the RTIM program operates by testing each cross-section on the diagram in turn to determine whether the balance line should be passed through it. This is done by searching ahead of the cross-section of interest to find the next cross-section at which the accumulated earthworks balance is the same. If the distance between these cross-sections is less than the marginal haul, a balance line is set up joining the two cross-sections and material will be hauled and balanced between them. The volume of material and the haul involved for this particular loop are found by calculating the area of the balance loop. If the distance between these cross-sections is greater than the marginal haul, then it will not be economical to haul material between them. The material between the cross-section of interest and the end of the previous balance line is assumed to be spoiled if it is cut, and borrowed if it is fill. In this procedure, the subsequent cross-section is then considered to see if an economical balance loop can be started there. The procedure continues until the end of the road has been reached.

Haulage costs are found by multiplying the total area of all the balance loops by the unit haulage rate—expressed as the cost of hauling one cubic metre of material one kilometre. The total borrow volume found by the mass-haul strategy is multiplied by the unit cost to give the total cost of borrow. The amount of unsuitable material found from the calculation of the suitable cut volume is added to the spoil volume by the mass-haul strategy to give the total volume of spoil. This total volume is multiplied by the unit cost to give the cost of spoil.

This elegant method gives a good level of accuracy and is very appropriate for determining the earthworks costs of alignments designed to different standards as part of a life cycle cost analysis.

4.4.2 Pavement design

Many methods exist for the design of asphalt pavements, including those developed by AASHTO (1974), Shell (1978) and TRRL (Powell et al., 1984). Similarly, for concrete pavements, design methods have also been produced by the Portland Cement Association (PCA, 1966) and TRRL (Mayhew and Harding, 1987). All methods of pavement design require knowledge about the traffic loading that the pavement will carry over its life.

The amount of damage done to a road by a moving vehicle depends very strongly on the axle load of the vehicle. The relationship between the damage and the axle load is extremely important for the proper design of pavements. To help with this design, 'equivalence factors' are normally used. The equivalence factor of a vehicle is defined as the number of passages of a 'standard' 80 kN axle which would do the same damage to a road as one passage of the vehicle in question (Liddle, 1963). The equivalence factor for each class of vehicle is calculated from the gross vehicle weight using the following relationship which is a simplified derivation of Liddle's formula:

$$EF = \sum_{n=1}^{k} \left[\frac{L_i}{80} \right]^n \qquad (4.4)$$

where L = load in kN on axle i, k = number of axles on the vehicle and n = a factor, normally in the region of 4.

Pavement design 'lives' are, therefore, normally expressed in terms of the number of equivalent 80 kN standard axles that the pavement can carry before structural 'failure' occurs. Failure, in this sense, is normally defined in terms of extent of cracking or amount of rutting, rather than as complete collapse of the pavement.

For life cycle costing purposes, the main aspects of pavement design that are of concern are the cost and the life under traffic before renewal is needed. Once the design has been carried out, costing is normally fairly straightforward, being a question of applying unit rates to the volumes of the different materials in each of the layers. The life will depend on the pavement strength, the traffic loading, and the resulting rate of deterioration. Pavement strength of asphalt roads is normally defined in terms of a 'modified structural number' (SNC), given by (Hodges et al., 1975):

$$SNC = 0.0394 \sum_{i=1}^{n} a_i d_i + 3.51 \log_{10} CBR$$

$$- 0.85(\log_{10} CBR)^2 - 1.43 \qquad (4.5)$$

where a_i = the AASHTO strength coefficient of layer i in the pavement—see Watanatada et al. (1987a) for typical values, d_i = the thickness of layer i in mm, n = the number of pavement layers and CBR = the in-situ California bearing ratio of the subgrade. Resulting rates of deterioration are discussed in the next section.

In addition to the cost of initial pavement construction, there is normally a need to cost pavement overlays or pavement reconstructions during the life cycle analysis period. This is to take account of the need to strengthen the pavement during the road's life or to provide extra capacity to carry increased volumes of traffic. In developing countries, consideration must be given to the economics of upgrading gravel roads to paved surfaces as traffic flows exceeds the threshold at which this becomes economic. In all these cases, providing that the dimensions and materials of the pavements are known, costing is relatively straightforward.

4.4.3 Drainage and structures

Many standard methods exist for the design of drainage facilities, bridges and other structures. These normally have a design life which is at least as long as any analysis period used for life cycle cost analysis. As such, their cost is unlikely to be influenced by factors that will change under different analysis scenarios, and is not considered further here.

4.4.4 Cost estimating

Costing for life cycle analysis has been based traditionally on a unit price analysis. In some cases, however, it has been found that cost estimates produced by this method can be unreliable. A report produced by UMIST (1987) found that the method was deficient in several important areas. It recommended, instead, the use of analytical techniques and rigorous procedures of risk management to produce realistic estimates of cost at all stages.

UMIST recommended that expected values of construction costs should reflect past experience from completed projects since actual values achieved have normally been far in excess of those estimated originally, particularly those estimates produced at the early stages of the design.

The UMIST approach to costing recognises that it is necessary to expend considerable time and effort at all stages of the project if realistic estimates of cost are to be produced, and allowance for this should be made when life cycle cost analysis is being carried out.

Four base estimating techniques are recommended at different stages of the project cycle:

1. Global, or 'broad brush', estimates.
2. Man-hours estimates used principally for works involving large amounts of labour.
3. Unit rates based on historical records of completed works.
4. Operational, or resource cost, estimates compiled from the fundamental consideration of the constituent operations or activities revealed by the method statement and programme, and from the accumulated demand for resources.

Also recommended is the inclusion in the estimates of separate allowances to cover contingencies. These are of two types:

1. Expected costs, which have not been separately identified, but which experience indicates must inevitably occur during construction, and can be covered by a lump sum or a percentage value.
2. Tolerances, based on past experience, which are an estimate of the probability of unforeseen costs arising and of their probable magnitude; these reflect the fact that costs may over run due to physical contingencies, such as unexpectedly poor ground conditions or lack of finance which prolongs construction time.

4.5 Deterioration and maintenance

4.5.1 Bitumen surfaced roads

Bituminous road pavements deteriorate over time under the combined effects of traffic and weather. The wheel loadings of heavy traffic induce levels of stress and strain within the pavement layers which are functions of the stiffness and layer thicknesses of the materials. Under repeated loadings, these cause the initiation of cracking through fatigue in bound materials and the deformation of all materials.

Weathering causes bituminous surfacing materials to become brittle and thus more susceptible to cracking and to disintegration, including ravelling, spalling and edge breaking. Once initiated cracking extends in area, increases in intensity (closer spacing) and increases in severity (width of crack) to the point where spalling and, ultimately, pot-holes develop. Open cracks on the surface and poorly maintained drainage systems permit excess water to enter the pavement. This hastens the process of disintegration, reducing the shear strength of unbound materials, and thus increasing the rate of deformation under the stresses induced by traffic loading.

The cumulative deformation throughout the pavement is manifested in the wheelpaths as ruts and, more generally, in the surface as an unevenness or distortion of the profile, which is termed roughness. Environmental effects of weather and seasonal changes then cause further distortions. The roughness of a pavement is, therefore, the result of a chain of distress mechanisms and the combination of various modes of distress. Maintenance is usually intended to reduce the rate of deterioration, but certain forms, such as patching, may even increase the roughness slightly. Roughness is thus viewed as a composite distress, comprising components of deformation due to traffic loading and rut depth variation, surface defects from spalled cracking, pot-holes, and patching, and a combination of ageing and environmental effects.

Table 4.1 Classification of pavement distress by mode and type

Mode	Type	Brief description
Cracking	Crocodile	Interconnected polygons of less than 300 mm diameter
	Longitudinal	Linear cracks along the length of the pavement
	Transverse	Linear cracks across the pavement
	Irregular	Unconnected cracks without a distinct pattern
	Map	Interconnected polygons more than about 300 mm in diameter
	Block	Intersecting linear cracks in rectangular pattern at spacing greater than about 1 m
Disintegration	Ravelling	Loss of stone particles from the surfacing
	Pot-holes	Open cavity in surfacing, greater than about 150 mm in diameter or than about 50 mm in depth
	Edge-break	Loss of fragments at the edge of the surfacing
Deformation	Rut	Longitudinal depression in wheelpaths
	Depression	Bowl-shaped depression in surfacing
	Mound	Localised rise in surfacing
	Ridge	Longitudinal rise in surfacing
	Corrugation	Transverse depressions at a spacing of less than about 5 m
	Undulation	Transverse depressions at a longer spacing
	Roughness	Irregularity of pavement surface in wheelpaths

Source: Paterson (1987).

The deterioration of bituminous roads can therefore be classified under the following modes of distress:

- cracking
- disintegration
- permanent deformation.

These are categorised further by distress types in Table 4.1.

The nature of pavement deterioration is that all distress modes and types interact with each other. In addition, all are affected by maintenance and this interaction is a key issue for life cycle costing. The ultimate effect of the interaction is manifest in levels of road roughness.

4.5.2 Roughness

The cost of operating vehicles and transporting goods rises as road roughness increases. Since the total operating costs of vehicles outweigh the highway authority's costs of road maintenance typically by a factor of between 10 and 20, small improvements in roughness can yield high economic returns. The economic impact of roughness is, therefore, considerable and provides the strongest objective basis for evaluating road policies when using life cycle analysis.

The returns from improved roughness are not immediately apparent to the highway authority, because most of the benefits accrue to the road users. However, it is the road users who also bear the costs of neglected maintenance.

Thus, the benefits are realised through lower transport costs and, ultimately, in the economy more widely.

For this reason, the focus here is on relationships for road roughness progression which are appropriate to life cycle cost analysis. The relationships developed by Paterson (1987) are the most comprehensive.

Roughness progression is predicted as the sum of three components:

1. Structural deformation, related to
   - roughness
   - equivalent standard axle flow
   - structural number.
2. Surface condition, related to changes in
   - cracking
   - pot-holing
   - rut depth variation.
3. Pavement age and environment-related roughness.

Paterson's relationships apply to all bituminous pavements. His relationship for the predicted incremental change in road roughness, in m/kmIRI (international roughness index), due to road deterioration during a year is:

$$\Delta R_d = 0.929\, K_{gp} \cdot F + 0.023\, K_{ge}(14 \cdot R_a - 0.714) + 0.714 \qquad (4.6)$$

where R_a = the roughness at the start of the year, in m/kmIRI*, K_{gp} = a user-specified deterioration factor for roughness progression (default value = 1), K_{ge} = a user-specified deterioration factor for the environment-related annual fractional increase in roughness (default value = 1), and F = contribution to roughness of structural deformation and surface condition given by:

$$F = 134\text{EMT}(\text{SNCK} + 1)^{-5.0}\text{YE4} + 0.114(\text{RDS}_b - \text{RDS}_a)$$
$$+ 0.0066\Delta\text{CRX}_d + 0.42\Delta\text{APOT}_d$$

where EMT = $\exp(0.023 K_{ge}\text{AGE3})$, AGE3 = the construction age, defined as the time since the last overlay, reconstruction or new construction activity, in years, SNCK = the modified structural number adjusted for the effect of cracking, and given by SNCK = max(1.5; SNC − ΔSNK), SNC = the modified structural number, as defined earlier, ΔSNK = the predicted reduction in structural number due to cracking since the last pavement reseal, overlay or reconstruction*, YE4 = the number of equivalent standard axle loads for the analysis year, based on an axle-load equivalency component of 4.0, in millions/year, RDS_b = the standard deviation of rut depth at the end of the year*, RDS_a = the standard deviation of rut depth at the beginning of the year*, ΔCRX_d = the predicted change in area of indexed cracking due to road deterioration*, ΔAPOT_d = the predicted change in the total area of pot-holes during the analysis year due to road deterioration, in %*. The derivation of definitions marked with an asterisk (*) will be found in Paterson (1987) or Watanatada et al. (1987a).

4.5.3   Concrete roads

The mechanism of deterioration of concrete roads are quite different to those of roads with a bituminous surface. Cracking is the dominant indicator of distress, but this tends to be caused by lack of sub-base support under the concrete slab, rather then by fatigue. Lack of support can be due to inadequate, or variable, compaction, or to settlement or leaching of subgrade materials. Linear cracking can be either longitudinal or transverse, or cracking can be multiple and interconnected, known as map cracking. Settlement of concrete pavements can also occur, where the whole slab tilts, but does not necessarily crack.

The surface of concrete pavements can deteriorate through fretting, where aggregates are plucked from the surface, or scaling, where thin slivers of material become removed. Both are caused by poor adhesion in the concrete matrix and are exacerbated by the freezing of water that may have been allowed to enter.

Much deterioration of concrete roads is associated with the joints between slabs. These are normally sealed and problems can arise where the sealant is allowed to deteriorate or become damaged. This inhibits expansion and contraction due to thermal changes, which results in damage to the pavement. Occasionally, whole slabs rock on the passing of traffic, with the result that underlying material, in the form of a slurry, is pumped up through the damaged joints.

The nature of concrete road deterioration, with its relationships to problems in the sub-base, subgrade and joints, means that it is difficult to model its deterioration in a satisfactory manner by using deterministic methods. Many early problems are, in fact, construction defects. As a result, the relationship between life cycle costs and the design, construction and subsequent deterioration of concrete roads is not well defined, and no relationships are presented here.

4.5.4   Unpaved roads

Three-quarters of the roads around the world do not have a paved surfacing; instead, they consist of rudimentary earth tracks or are covered with gravel to provide resistance to the wear and tear of traffic and rainfall. Although these roads carry relatively low volumes of traffic, many play a key role in the economic and social activity of areas in developing countries. They often provide the cheapest means of transporting agricultural produce to markets, and the only access to settlements for the provision of medical and agricultural extension services.

Deterioration of unpaved roads is manifest principally as rutting and roughness. In certain cases, roughness can take the form of 'corrugations' (Heath and Robinson, 1980), which are regular undulations typically of 0.5–1.0-m spacing and with depths often of several centimetres. These provide a very uncomfortable ride and are very damaging to vehicles using the road. In road surfacing materials that are prone to corrugations, formation of the undulations often recurs within a few days of maintenance having been carried out. They are a particular problem in dry areas. A further problem in arid zones is that loose, fine material covers the road surface, causes drag on the traction of vehicles and causes clouds of dust which makes driving conditions both inefficient and dangerous. All gravel roads deteriorate as a result of material that is lost from the road surface due to abrasion by tyres and the effects of the elements (Jones, 1984).

For the purposes of life cycle costing, the key deterioration factors are roughness and gravel loss: roughness because of its effect on vehicle operating costs, and gravel loss because of the high cost of replacing lost material at intervals through the road's life. A key use of life cycle costing for unpaved roads is to determine the optimum timing when it is economic to provide a paved surfacing. This is typically when traffic levels reach between 200 and 400 vehicles per day, but the actual figure depends very much on local conditions, on the costs of maintenance and on the availability of suitable materials for regravelling.

The most comprehensive unpaved-road deterioration relationships are probably those developed by Jones (1987), which are quoted here. Those developed by Paterson (1987) should also be considered.

Roughness

$$R = 0.0032(5153 + 74.631T - 39.17P075 - 82.534PI$$

$$- 688.57R_L - 18.315VC)^{0.89} \qquad (4.7)$$

where R = roughness, in m/kmIRI, T = traffic in both directions since maintenance blading, in thousands of vehicles, $P075$ = percentage of gravel surfacing material passing the 0.075 mm sieve, PI = plasticity index of wearing course, in %, R_L = annual rainfall, in metres and VC = vertical gradient of the road, in m/km.

Gravel loss

$$GL = 11.276 + 0.00325\,ADT \cdot P075 - 0.293PI + 2.662R$$

$$+ 0.1542VC \qquad (4.8)$$

where GL = annual gravel loss, in mm, ADT = annual average daily traffic (in both directions) and other variables are as above.

4.5.5 Maintenance

Deterioration can be arrested by the implementation of maintenance. The effect on deterioration will depend on the nature of the maintenance, its

Table 4.2 Maintenance for bituminous surfaced roads

Class	Frequency	Treatment options	Effects
Routine	Annual	Maintenance of drainage, road furniture, etc.	Indirect (absence would lead to other defects)
Patching	Periodic (intervals of greater than one year)	Surface patching, local sealing, deep (base) patching	Prevents local defects becoming more widespread; eliminates local cracking and deformation and pot-holes
Resealing	Periodic	Fog seal/surface rejuvenation, slurry seal, surface dressing, resurfacing	Increases life before strengthening is needed by sealing cracks, correcting other surface defects, and renewing waterproofing
Strengthening	Periodic	Overlay on existing surface, replacement and reconstruction of surface	Provides pavement with new design life by eliminating cracking and other surface defects and deformation, and reducing roughness

relevance to the defect being corrected, and the timeliness of its intervention. For bituminous surfaced roads, the maintenance treatments shown in Table 4.2 are typical of those used to correct defects.

Maintenance of concrete roads consists of cleaning and resealing joints on a routine basis and attention to the maintenance of drainage and the like, as for bitumen surfaced roads. There is no concept of periodic maintenance to the road surface and structure in the sense that it can be planned in advance in a similar way to that for bitumen surfaced roads; this maintenance is responsive to the appearance of defects. Cracks are sealed and scaled areas are patched. These activities are undertaken using bitumen, cement grout or epoxy. Major problems with slabs are normally corrected by breaking out the affected slabs and replacing them in their entirety, although, where slabs remain intact, it is possible to jack up the slab and to pressure-grout the void underneath. Where joints have seized or become damaged, then the part of the slab immediately adjacent to the joint must be replaced.

Unpaved roads are maintained principally by smoothing out rough surfaces and by replacing gravel surfacing material that is worn away. Smoothing and reshaping is normally undertaken by a blade grader although, as an intermediate operation, this is sometimes undertaken by towing simple drags along the road to remove loose material. Regravelling is essentially a surface reconstruction operation, by adding new material and spreading and compacting it to form a new surface. Local (spot) regravelling is sometimes undertaken to treat specific problem areas, and routine maintenance is carried out to the off-carriageway features in a manner similar to bitumen and concrete roads.

4.6 Road user costs

4.6.1 Vehicle operating costs

When a road improvement is undertaken, the owners and users of vehicles profit from reduced costs of transport. Higher average speeds can be maintained, and the more even running, with fewer gear changes and less braking, may lead to savings in fuel consumption. Tyres last longer on improved road surfaces, and there is less wear and tear on the suspension and body. These savings are perceived by vehicle operators in the form of lower expenditures.

Vehicle operating costs depend on the number and types of vehicles using the road, the road curvature, gradient and road width, the roughness of the surface of the road, and driver behaviour. Changes in any of these parameters as a result of a road improvement will result in a change in vehicle operating cost.

The components of vehicle operating cost, with their approximate respective contributions to the total, are given in Table 4.3. This is based on a sample of feasibility studies carried out by consultants as part of the field testing of the prototype RTIM life cycle costing program (Robinson, 1976b). This table gives an indication of the relative importance of individual variables when making estimates of vehicle operating costs.

Vehicle depreciation and overheads have been shown effectively to be independent of life cycle highway cost considerations; accordingly, the costs discussed further here will be those for fuel consumption and spare parts,

Table 4.3 Relative contribution of vehicle operating cost components

Component	Percentage contribution	
	Private cars	Trucks
Fuel consumption	10–35	10–30
Lubrication oil consumption	< 2	< 2
Spare parts consumption	10–40	10–30
Vehicle maintenance labour hours	< 6	< 8
Tyre consumption	5–10	5–15
Vehicle depreciation	15–40	10–40
Crew costs	0	5–50
Other costs and overheads	10–15	5–20

Source: Robinson (1988b).

which are road roughness-dependent and make a significant contribution to vehicle operating costs.

4.6.2 Fuel consumption

The key parameter that affects fuel consumption is vehicle speed. Although current speed values are best determined by direct measurements of traffic on the road, it is necessary to predict future changes in speed as a result of changing road conditions or possible traffic congestion.

The World Bank's method (Watanatada et al., 1987a, b) uses what is described as an 'aggregate limiting velocity approach' to steady-state speed prediction. This method works on the assumption that speed is limited by several independent parameters related to the vehicle and to road condition, including gradient, engine power, road curvature and road roughness. Speed is estimated when constrained by each of the parameters, independently, in turn. The actual predicted speed is then determined as the probabilistic minimum of those constraining speeds. The constraining speeds used are (in m/s):

> VDRIVE = the limiting speed based on vertical gradient and engine power
> VBRAKE = the limiting speed based on vertical gradient and braking capacity
> VCURVE = the limiting speed determined by road curvature
> VROUGH = the limiting speed based on road roughness and associated ride severity
> VDESIR = the desired speed in the absence of other constraints based on psychological, economic, safety and other considerations.

Using the respective values of the five limiting speeds for each road segment, the predicted steady state speed for the segment is computed from:

$$V_x = \frac{E_0}{\left[\left(\frac{1}{\text{VDRIVE}_x}\right)^{1/B} + \left(\frac{1}{\text{VBRAKE}_x}\right)^{1/B} + \left(\frac{1}{\text{VCURVE}}\right)^{1/B} + \left(\frac{1}{\text{VROUGH}}\right)^{1/B} + \left(\frac{1}{\text{VDESIR}}\right)^{1/B}\right]^B} \quad (4.9)$$

where E_0 = is a factor for correcting bias arising from the nonlinear transformation of the variables used in the estimation process, B = is a coefficient which determines the shape of the probability distribution and x = the uphill and downhill segments, recognising that there is a gravity effect on certain of the parameters. The determination of the constraining speeds, VDRIVE, VBRAKE, VCURVE, VROUGH and VDESIR, is given by Watanatada et al. (1987a).

For a vehicle operating on any road section of specified geometric alignment, the average round trip fuel consumption, FL, in litres/1000 vehicle-km is given by:

$$FL = 500\alpha_1\alpha_2 \left[\frac{UFC_u}{V_u} + \frac{UFC_d}{V_d} \right] \qquad (4.10)$$

where α_1 = constraint to reflect the energy efficiency of vehicles in different countries (default = 1.0), α_2 = constraint to reflect the difference between the experimental conditions under which the relationships were derived and real-life conditions (default = 1.15), V_u = speed in the uphill direction, from equation (4.9), V_d = speed in the downhill direction, from equation (4.9),

$$UFC_u = [UFC_0 + a_3 HP_u + a_4 HP_u CRPM + a_5 HP_u^2] \times 10^{-5}$$

$$UFC_d = \begin{cases} [UFC_0 + a_3 HP_d + a_4 HP_d CRPM + a_5 HP_d^2] \times 10^{-5} & \text{if } HP_d \geqslant 0 \\ [UFC_0 + a_6 HP_d + a_7 HP_d^2] \times 10^{-5} & \text{if } NH_0 \leqslant HP_d < 0 \\ [UFC_0 + a_6 NH_0 + a_7 NH_0^2] \times 10^{-5} & \text{if } HP_d > NH, \end{cases}$$

where $UFC_0 = a_0 + [a_1 + a_2] CRPM$, CRPM = the calibrated engine speed, in revolutions per minute, HP_u, HP_d = the vehicle powers on the uphill and downhill road segments, in metric horse-power and a_i = constant whose value depends on vehicle type. These relationships apply to free-flow traffic conditions. When congestion becomes a factor, speed-flow relationships should be used, and these are discussed later.

4.6.3 Spare parts consumption

It is most convenient to model spare parts consumption in terms of the ratio of the monetary cost of the parts consumed per 1000 vehicle-kilometres to the price of a new vehicle in the same period. This overcomes the problem of estimating the cost of all individual parts contained in individual vehicles.

The parts cost per 1000 vehicle-kilometres for the given vehicle class expressed as a fraction of the cost of a new vehicle is given by:

$$PC = \begin{cases} CKM^k C_{osp} \exp[C_{spqi}(14IRI - 10)] & \text{for IRI} < IRI_{osp} \\ CKM^k [a_0 + a_1(14IRI - 10)] & \text{for IRI} > IRI_{osp} \end{cases}$$

where CKM = the average age of the vehicle group in km, defined as the average number of kilometres that the vehicles have driven since new, k = an exponent, depending on the vehicle type, C_{osp} = a constant, C_{spqi} = a constant and IRI_{osp} = the transitional value for roughness, in m/km IRI, beyond which the relationship between spare parts consumption and roughness is linear. More detailed explanation of the constants is given by Watanatada *et al.* (1987*a*).

4.6.4   *Time savings*

Journey time savings can represent a large proportion of a project's benefit, and hence can be of great significance in life cycle cost analysis. The benefits of shorter journey times will accrue to the vehicle fleet, in that greater vehicle productivity can be achieved, and to the passengers and freight being carried.

It has been argued that increased vehicle utilisation cannot occur as a result of route shortening because journeys are 'quantised', and only savings that are large enough to permit an extra trip should be counted as a benefit. However, although the vehicle fleet cannot be employed for 100% of the working day, it is reasonable to assume that, after the project is completed, vehicles work, on average, for the same proportion of the working day as in the 'without project' situation. There is no reason for assuming that there is some special feature of the 'before' situation that will not be matched after the project is completed, and hence it is reasonable to assume that potential time savings will always be fully utilised over the network as a whole. It is therefore appropriate that fleet savings due to utilisation are included as a benefit in life cycle cost analysis.

Values of time can be based on the approach adopted in the UK (Department of Transport, 1981), although the actual unit time values used will depend on incomes in the country where the life cycle cost analysis is to be carried out. The following issues should be considered:

1. Time savings should be measured separately for working and leisure time.
2. In the absence of better data, working time should be valued at the average wage rate in the monetised economy of the country, or the relevant region within it, plus overheads.
3. Non-working or leisure time should be valued in the range 0–45% (the value used in the UK) of working time, unless there are special reasons for attributing a higher value; the value within this range will probably be related directly to the GNP per capita of the country.

Time costs for freight comprise the costs of interest on capital which the goods represent, costs due to damage or spoilage of perishable goods, and ancillary costs which might arise because of delay. Studies of modal choice for goods travelling by road or other mode have suggested that, even for non-perishable goods, consignors are usually willing to pay far more than just interest cost on the goods to reduce travel time, and are more concerned with the reduction of uncertainty in the time of delivery.

The key to evaluating time savings in a life cycle analysis is the prediction of travel speed changes over the life cycle of the road. In situations where congestion is not a significant factor, the speed relationships in the earlier section on fuel consumption may be used. In those situations where congestion will have a significant effect, speed-flow curves should be used.

A speed prediction formula, which can be applied to rural single-carriageway roads which are subject to congestion, is used in the COBA9 method (Department of Transport, 1981).

$$VL = 76.5 + 1.1CW + VISI/60 - 5(VF + AF) - 2B/45 \qquad (4.11)$$
$$- 5H/45 - 2.5NG/45 - SLS \cdot Q$$
$$VH = 83.5 + VISI/120 - 5(VF + AF) - 2B/45 - 9H/45 \qquad (4.12)$$
$$- 6NG/45 - SHS \cdot Q$$

where VL = speed of light vehicles (km/h), VH = speed of heavy vehicles (km/h), CW = average carriageway width (m), $VISI$ = average sight distance (harmonic mean) (m) given by:

$$VISI = n/[(1/x_1) + (1/x_2) + \cdots + (1/x_n)]$$

where n = number of observations, x_i = sight distance at point i, VF = verge 'friction' = $2(VW + 1)$, VW = average verge width, both sides, including hard strips (m), AF = access 'friction' = AXS/CW, AXS = total number, both sides, of lay-bys, side roads and accesses excluding houses and field entrances (number/km), B = bendiness: total change of direction per unit distance (degree/km), H = hilliness: total rise and fall per unit distance (m/km), (Figure 4.3), NG = nett gradient: nett rise per unit distance on one-way links (m/km), Q = total flow of all vehicles per standard lane (vehicles per hour per 3.65-m lane), SLS = the speed/flow slope for light vehicles ($Q \leqslant 1200$), $SLS = (21 + D \cdot P/100)/1200$, $D = VL - VH$ at $Q = 0$, P = percentage of heavy vehicles, SHS = the speed/flow slope for heavy vehicles ($Q \leqslant 1200$), $SHS = [21 - D(100 - P)/100]/1200$ and $SLS = SHS = 21/1200$ for $Q > 1200$.

The speed/flow slopes are defined to ensure that $VL = VH$ at $Q = 1200$ veh/h per standard lane. This constraint is imposed because, at high flows on a single carriageway road, it is impossible for one class of vehicle to travel at consistently different speeds from another.

The speed prediction formula (Department of Transport, 1981) for light vehicles on dual carriageway roads is as follows. Where HR = sum of rises per unit distance (m/km) and HF = sum of falls per unit distance (m/km), then:

   (i) If HR and HF do not exceed 40 m/km

$$VL = C - (B/10) + (HF/4) - SLD \cdot Q \qquad (4.13)$$

where $C = 108$ for dual 3-lane motorways, 104 for dual 2-lane motorways, 103 for dual 3-lane all-purpose roads and 98 for dual 2-lane all-purpose roads.

   (ii) If $HF > 40$, then HF is set to 40 in the formula

   (iii) If $HR > 40$, then $(HR - 40)/2$ is subtracted from the formula.

For heavy vehicles on dual carriageway roads the speed prediction is as follows:

(i) For motorways, if $HR \leqslant 20\,m/km$

$$VH = 93 - (B/10) - HR - SHD \cdot Q \qquad (4.14)$$

(ii) For all-purpose roads and motorways, if $HR < 20\,m/km$

$$VH = 83 - (B/10) - (HR/2) - SH \cdot Q \qquad (4.15)$$

The speed/flow slopes for light and heavy vehicles are, respectively:

$$SLD = 6/1000 \quad \text{for } Q < 1200\,veh/h$$
$$= 27/1000 \quad \text{for } Q > 1200\,veh/h$$
$$SHD = 0 \quad \text{for } Q < 1200\,veh/h$$
$$= 14/1000 \quad \text{for } Q > 1200\,veh/h$$

COBA9 also contains speed prediction relationships for urban conditions.

4.6.5 Reduction in road accidents

In order to determine road accidents benefits, it is necessary to:

- forecast the reduction in accidents
- determine the appropriate values for the costs of accidents.

Methods of forecasting road accident reduction are still fairly subjective and there is a general lack of data about the effects of various remedial measures on accident rates. It is important that a distinction should be drawn between:

- accident prevention resulting from improved standards of highway design and planning
- accident reduction resulting from engineering countermeasures introduced to improve road safety at specific sites.

The costs of road accidents have three components:

- damage to vehicles and other property
- costs of police work, hospital treatment, administration, etc.
- loss of life and injury.

Losses such as damage, police and hospital costs, involve material resources and are defined readily, even though the appropriate value may be subject to conjecture. However, costs relating to loss of life and injury are subjective, involving the need to value human life and 'pain, grief and suffering'. Material costs should always be included to provide an absolute minimum value for accident reduction. The appropriate method of valuing loss of life and injury in a particular country will depend on the road safety objectives of the government of that country.

If maximising GNP is an objective, then a costing method based on 'gross output' is appropriate. In this case, the value of human life is taken as the discounted present value of the casualty's future output. This assumes that there is no-one else able to do the casualty's work who would otherwise be unemployed. A component for pain, grief and suffering should be added to this if account is also to be taken to society's aversion to death and injury.

4.6.6 Wider economic benefits

In life cycle cost analysis of road projects, it is normal to treat generated traffic as a surrogate for economic benefits (Van der Tak and Ray, 1971). The 'consumer surplus' method of assessing developmental benefits is appropriate to most road projects because it is usual that some kind of vehicle access already exists, however rudimentary. Use of the consumer surplus approach to life cycle analysis of road projects is described by Robinson (1988b).

4.7 Consequences

The use of life cycle costing in the appraisal of highway projects ensures that account is taken of the medium to long-term consequences of decisions made now, rather than consideration only of short-term issues.

If life cycle costs are not taken into account, then minimum construction-cost solutions will always be chosen, constrained only by any design standards that might be applied. Design standards in common use have often been derived from considerations of custom and practice, and usually provide only a minimum level of safety and engineering functionally. Such an approach fails to recognise that the only reason for constructing a highway is to provide a service over a period of time into the future. An appraisal philosophy which fails to recognise this is clearly flawed.

A life cycle cost approach is appropriate for highway agencies to assist in planning programmes of maintenance. When maintenance engineers are faced with a broad array of defects over the network that they are managing, choices must be made about appropriate treatments for individual defects, and about the balance of treatments over the network as a whole, given likely constraints on available budgets. Only by considering life cycle costs can programmes of works be selected, both for individual sites and for the network as a whole, that minimise costs or maximise the use of available funds and ensure that optimum conditions prevail on the network both now and in the future.

The relevance of life cycle costing of highways to broader economic issues was highlighted by the studies first reported in the late 1970s which demonstrated for the first time, in a quantitative way, the key importance that road deterioration had on vehicle operating costs. This, in turn, highlighted the dramatic effect that spending, or failing to spend, relatively small amounts of money on road maintenance could have on the economic return of a road project. It became clear that, because road transport costs represented a significant portion of the GNP in many countries, the life cycle implications of road maintenance had considerable significance.

An example has been quoted (Robinson, 1988a) where a 100-km length of road in a developing country had received no maintenance in the four years that it had been open to traffic. The road, which was carrying approximately 750 vehicles per day, was already cracked and deformed, and pot-holes were starting to appear. This deteriorated road condition was already leading to an extra vehicle operating cost estimated to be about $1.5 million per year. Life cycle cost analysis showed a dramatic benefit from applying relatively low-cost maintenance measures.

More generally, life cycle cost analysis of road schemes in developing countries has led to important policy changes from organisations such as the World Bank. On roads that are currently maintained poorly, improvements in maintenance reduce vehicle operating costs typically by 15–50% for the same traffic level, with the result that it is common for internal rates of return on such projects to exceed 100%. Few maintenance projects, in such situations, have rates of return as low as 50%, while the return on new construction projects rarely exceeds this figure. The use of life cycle analysis to demonstrate such results has led some international funding agencies to switch lending more in favour of road maintenance projects and away from the traditional area of new construction.

The results of using life cycle cost analysis to appraise road schemes in developing countries have been evaluated by White (1984), with the following conclusions:

1. Roads with low baseline flows:
   - benefits from upgrading an existing track are likely to be small
   - benefits from providing completely new access, or a road improvement leading to a change in mode, are likely to be substantial, but difficult to measure
   - personal travel is likely to be most significant component of rural traffic flows.
2. Roads with high baseline flows:
   - high-income groups are likely to benefit more than low income groups
   - time savings are a main source of project benefit from major road projects, even when time values are low.
3. Comparison of different options using a common life cycle costing methodology is essential; it is insufficient to appraise in detail a pre-selected project.

4. Forecasting future traffic flows is subject to much greater errors in developing countries than in industrialised countries.
5. Good quality control at construction is very important when life cycle costs are being taken into account.
6. The level of maintenance throughout the life of a project has a significant impact on life cycle costs.
7. Lower design standards on rural access roads will improve the net present value.

Life cycle cost analysis plays an important part in the decision-making process in the highways subsector. This importance will continue to increase as interrelationships and analysis models evolve in the future.

Acknowledgement

Figures 4.4, 4.5 and 4.6 are published with the permission of the Chief Executive of the Transport Research Laboratory.

References

AASHTO (1974) AASHTO interim guide for design of pavement structures 1972, American Association of State Highway and Transportation Officials, Washington, DC.
Abaynayaka, S.W. *et al.* (1976) Tables for estimating vehicle operating costs on rural roads in developing countries, TRRL Laboratory Report LR 723, Transport and Road Research Laboratory, Crowthorne, UK.
Abaynayaka, S.W. *et al.* (1977) Prediction of road construction and vehicle operating costs in developing countries, *Proceedings of the Institution of Civil Engineers*, **62**(1), 419–446.
Abell, R. (1992) Whole life costing of road pavements, in: IHT Alan Brant National Workshop, Royal Spa Centre, Leamington Spa, 14 April 1992, Institution of Highways and Transportation, London, pp. 103–110.
Abell, R. *et al.* (1986) Estimation of life cycle costs of pavements, in: *Proc. International Conference on Bearing Capacity of Roads and Airfields*, Plymouth 16–18 September 1986, WDM, Bristol.
Barber, V.C. *et al.* (1978) The deterioration and reliability of pavements, Technical Report S-78-8, US Army Engineer Waterways Experimental Station, Vicksburg, Mississippi.
Both, G.I. and Bayley, C. (1976) Evaluation procedures for rural road and structure projects, in: *Proc. 8th Conference of the Australian Road Research Board*, Perth 23–27 August 1976, Part 6, Volume 8, Session 29, Australian Road Research Board, Nunawading, pp. 6–25.
Chesher, A. and Harrison, R. (1987) *Vehicle Operating Costs: Evidence from Developing Countries*, Johns Hopkins for the World Bank, Baltimore.
Croney, D. (1977) *The Design and Performance of Road Pavements*, HMSO, London.
CRRI (1982) Road user cost study in India: final report, Central Road Research Institute, New Delhi.
Davies, H.E.H. (1972) Optimizing highway vertical alignments to minimize construction costs: program MINERVA, TRRL Laboratory Report 463, Transport and Road Research Laboratory, Crowthorne, UK.
Department of Transport (1981) *COBA9 Manual*, Assessment Policy and Methods Division of the Department of Transport, London.
Feighan, K.J. *et al.* (1981) Application of dynamic programming and other mathematical techniques to pavement management systems, *Transportation Research Record*, **1200**, 90–98
Ferry, D.J.O. and Flanagan, R. (1991) Life cycle costing—a radical approach, CIRIA Report

No. 122, Construction Industry Research and Information Association, London.

Findakly, H. et al. (1974) Stochastic model for analysis of pavement studies, *Journal of Transportation Engineering, American Society of Civil Engineers*, **100**, Paper 10358, 57–80

Garcia-Diaz, A. et al. (1981) *Computerized Method of Projecting Rehabilitation and Maintenance Requirements due to Vehicle Loadings*, Texas Transportation Institute, Texas A&M University, Austin, Texas.

Garrett, C. (1985) Whole-life costing of roads, *Municipal Engineer*, **2** (August), 223–232.

Harral, C.G. et al. (1979) The highway design and maintenance standards model (HDM): model structure, empirical foundations and applications, in: *Proc. PTRC Summer Annual Meeting*, 9–12 July, PTRC, London.

Harrison, R. and Chesher A.D. (1983) Vehicle operating costs in Brazil: results of road user survey, *Transportation Research Record*, **898**, 365–373.

Heath, W. and Robinson, R. (1980) Review of published research into the formation of corrugations on unpaved roads, TRRL Supplementary Report 610, Transport and Road Research Laboratory, Crowthorne, UK.

Hide, H. (1982) Vehicle operating costs in the Caribbean: results of a survey of vehicle operators, TRRL Laboratory Report 1031, Transport and Road Research Laboratory, Crowthorne, UK.

Hodges, J.W. et al. (1975) The Kenya road transport cost-study: research on road deterioration, TRRL Laboratory Report 673, Transport and Road Research Laboratory, Crowthorne, UK.

Howe, J.D.G.F. (1972) A review of rural traffic-counting methods in developing countries, RRL Report LR 427, Road Research Laboratory, Crowthorne, UK.

Howe, J.D.G.F. (1973) The sensitivity to traffic estimates of road planning in developing countries, TRRL Report LR516, Transport and Road Research Laboratory, Crowthorne, UK.

Jones, T.E. (1984) Dust emissions from unpaved roads in Kenya, TRRL Laboratory Report 1110, Transport and Road Research Laboratory, Crowthorne, UK.

Jones, T.E. (1987) Optimum maintenance strategies for unpaved roads in Kenya, Thesis submitted to the University of Birmingham for the degree of Doctor of Philosophy, University of Birmingham, UK.

Jung, F.W. (1985) Annual worth cost analysis of pavement rehabilitation, in: *Proc. Roads and Transportation Association of Canada 1985 Annual Conference*, Vol. 1, Roads and Transportation Association of Canada, Ottawa, pp. 173–197.

Kilbourn, P. and Abell, R. (1988) Whole life costs of concrete pavements, in: *Proc. PTRC Summer Annual Meeting*, University of Bath, 12–16 September 1988, PTRC, London.

Liddle, W.J. (1963) Application of AASHO Road Test result to the design of flexible pavement structures, in: *Proc. International Conference on the Structural Design of Asphalt Pavements*, University of Michigan, Ann Arbor, pp. 42–51.

Lindow, E.S. (1978) *Systems Approach to Life-cycle Design of Pavements*, Army Construction Engineering Research Laboratory, Champaign, Illinois.

Litten, M. and Johnston, B. (1979) *A Pavement Design and Management System for Forest Service Roads—Implementation*, US Forest Service, Washington, DC.

Loong, K.Y. (1989) Road pavement design and maintenance — a life cycle cost approach, in: *Proc. Technology Exchange Centre International Conference and Exhibition on Road Transport*, China Ministry of Communications, Beijing, pp. 195–204.

Markow, M.J. (1984) *EAROMAR*, Version 2, CMT Inc. for Federal Highway Administration, Cambridge, Massachusetts.

Mayhew, H.C. Harding, H.M. (1987) Thickness design of concrete roads, TRRL Research Report 87, Transport and Road Research Laboratory, Crowthorne, UK.

Moavenzadeh, F. (1972) Investment strategies for developing areas: analytical models for choice of strategies in highway transportation, Department of Civil Engineering Research Report No. 72–62, Massachusetts Institute of Technology, Cambridge, Massachusetts.

Moavenzadeh, F. et al. (1971) Highway design study phase I: the model, Economics Department Working Paper No 96, International Bank for Reconstruction and Development, Washington, DC.

Morosiuk, G. and Abaynayaka S.W. (1982) Vehicle operating costs in the Caribbean: an experimental study of vehicle performance, TRRL Laboratory Report 1056, Transport and Road Research Laboratory, Crowthorne, UK.

Ockwell, A. (1990) Pavement management: development of a life cycle costing technique, Bureau of Transport and Communication Occasional Paper 100, Australian Government Publishing Service, Canberra.

Parsley, L.L. and Robinson, R. (1982) The TRRL road investment model for developing countries, TRRL Laboratory Report 1057, Transport and Road Research Laboratory, Crowthorne, UK.

Paterson, W.D.O. (1987) *Road Deterioration and Maintenance Effects: Models for Planning and Management*, Johns Hopkins for the World Bank, Baltimore.

PCA. (1986) Thickness design for concrete pavements, Portland Cement Association, Washington, DC..

Petts, R.C. and Brooks J. (1986) The World Bank's HDM-III whole life cost model and its possible applications. in: *Proc. PTRC Summer Annual Meeting*, University of Sussex, 14-17 July 1986, PTRC, London; pp. 91-100.

Powell, W.D. *et al.* (1984) The structural design of bituminous roads, TRRL Laboratory Report 1132, Transport and Road Research Laboratory, Crowthorne, UK.

Rada, G.R. *et al.* (1986) Integrated model for project level management of flexible pavements, *Journal of Transportation Engineering, American Society of Civil Engineers*, **112** (4), 381–399.

Road Research Laboratory (1969) The cost of constructing and maintaining flexible and concrete pavements over 50 years, RRL Report LR 256, Road Research Laboratory, Crowthorne, UK.

Robinson, R. (1976a) Automatic generation of the highway vertical alignment: program VENUS, TRRL Laboratory Report 700, Transport and Road Research Laboratory, Crowthorne, UK.

Robinson, R. (1976b) The road transport investment model for developing countries: case studies, in *Proc. TRRL One-day Symposium on the Modelling of Total Costs of Road Transport in Developing Countries*, TRRL, 5 May 1976, Transport and Road Research Laboratory, Crowthorne, UK.

Robinson, R. (1988a) A view of road maintenance economics, policy and management in developing countries, TRRL Research Report 145, Transport and Road Resarch Laboratory, Crowthorne, UK.

Robinson, R. (ed.) (1988b) A guide to road project appraisal, TRRL Overseas Road Note 5, Transport and Road Research Laboratory, Crowthorne, UK.

Robinson, R. *et al.* (1975) A road transport investment model for developing countries, TRRL Laboratory Report 674, Transport and Road Research Laboratory, Crowthorne, UK.

Shell (1978) *Shell Pavement Design Manual: Asphalt Pavement and Overlays for Road Traffic*, Shell International Petroleum Company, London.

Stark, D.C. (1990) Appraisal of road schemes under conditions of suppressed demand, in: *Proc. PTRC Highway Appraisal Design and Management Seminar J. Summer Annual Meeting*, PTRC Education and Research Services, London, pp. 95-99.

Sullivan, T. and Scott, R. (1990) Strategic road network management--an approach using roughness, in: *Proc. 15th ARPB Conference*, Darwin, 26-31 August 1990, Australian Road Research Board, Vermont South Australia, pp. 105-113.

UMIST (1987) Construction project cost estimating study, Report for the Overseas Development Administration, University of Manchester Institute of Science and Technology, ODA, London.

Van der Tak, H.G. and Ray, A. (1971) The economic appraisal of road transport projects, World Bank Staff Occasional Papers Number Thirteen, Johns Hopkins Press, Baltimore.

Watanatada, T. *et al.* (1987a) *The Highway Design and Maintenance Standards Model*, Vols 1 and 2, Johns Hopkins for the World Bank, Baltimore.

Watanatada, T. *et al.* (1987b) *Vehicle Speeds and Operating Costs: Models for Road Planning and Management*. Johns Hopkins for the World Bank, Baltimore.

White, J.M. (1984) The evaluation of aid projects and programmes, in: *Proc. Overseas Development Administration Conference*, Institute of Development Studies, 7-8 April 1983, HMSO, London, pp. 64-75.

5 Life cycle costing in the defence industry
M.J. KINCH

5.1 Introduction

Life cycle costing is becoming established as an important tool of management in the defence industry. Costing models and systems have been developed independently by the controllerates which deal with sea and air defence systems, and are currently being considered by the land system controllerate. Although they have been used on relatively few equipments so far, the intention is to apply the techniques to all areas of defence procurement and system management. Eventually the management of life cycle costing will be brought under central direction.

This chapter is written with the intention of examining the benefits these techniques will be expected to confer on the economics of defence procurement once their use has become established, together with how they are likely to be employed, plus the problems that may have to be overcome, rather than reporting on the state of their development at present.

While life cycle costing techniques will be applied to the management of all defence equipment for use by all forces on land, sea and air, the principal examples used to illustrate points as they arise are taken from the air side. This is because aviation, military and civil alike, is a very strictly regulated activity, and a particular benefit of this is that the data necessary for the accurate management and monitoring of in-service support costs are much more readily accessible than they are for land and sea systems. Therefore support costs for future air systems can be forecast with greater confidence and with a higher likelihood of success. The importance of this will become clear as the chapter develops.

There are three main and interrelated reasons why life cycle costing is appropriate for the business of defence procurement and operation today. The first is the amount of money at stake compared with that in the past. The unit price of a World War 2 Spitfire was about £30 000; today a Tornado costs about £10 000 000; the EH 101 anti-submarine helicopter is priced at £24 000 000. In the USA the B-2 stealth bomber may have the distinction of becoming the first billion-dollar military aircraft.

The second lies in the length of time a weapon system may be expected to remain in service, and therefore the length of time that the quantitative arguments which governed the original decision to buy it have to remain valid. During the 1940s and 1950s a new type of aircraft was introduced into service every three years or so. The attritions of war and the rate of introduction of new developments in performance and weapons capability combined to make existing designs obsolete rapidly, and few of them were ever intended to last. Today planned lives of at least 25 years are not uncommon. The RAF's Canberras have been in use since 1950; the first Buccaneer flew in 1958 and the type was still active during the Gulf War in 1991; the Sea King has been in service since the early 1970s and will see the century out. A modern airframe has to be able to survive all the age-related depredations such as metal fatigue and corrosion that its predecessors never lasted long enough to encounter, while throughout its life it must remain capable of carrying successive generations of avionic equipment, much of which had never been even imagined at the time of its design. Both of these considerations lead to the initial cost of acquisition being very high.

The third reason arises from the first two. It is that the longer the service life is extended, the greater becomes the proportion of the whole life cost which is attributable to scheduled and unscheduled support. Therefore the earlier and the more successfully these costs can be forecast, the better the procurement decision will be.

Support costs, and especially those due to unreliability of current equipments, are currently causing the UK Ministry of Defence grave concern. For the year 1989–90 it was estimated that the costs associated with unexpected (i.e. unscheduled) maintenance of defence equipment exceeded £1 billion. This is the cost of rectification; the cost of the loss of availability while the defective equipment is out of service is one order of magnitude higher. Indeed, for the same period the Ministry calculated that unrealiability caused one-third to one-half of the RAF's fast jet fleet to be unavailable, and, even on those aircraft which could be operated, had a detrimental effect on one mission in ten.

Figures such as these clearly indicate the returns that could result from applying a wiser method of control of investment, especially where reliability and maintainability factors are concerned. However, the most important feature to recognise is that the cost figure that matters is not just the purchase price, as was thought to be the case about thirty years ago, but the costs of specification, development, purchase, operation and support all added together.

5.2 Optimal solution to a design requirement

Life cycle costing in defence applications may be defined as 'the technique of examining all the costs—in money terms—direct and indirect, social and environmental, of operating an equipment throughout its entire service life, as an aid to finding the optimal solution to a design requirement'. While this definition hints, not unjustly, that the figures will never be proved right until the equipment has reached the scrap-heap, its last few words hold the key to understanding the principal role that life cycle costing can play in defence procurement. This is that a common factor of cost will be imposed on all those variables that it was once easier to regard as unquantifiable.

5.2.1 Procedures

The first area in which life cycle costing techniques are effective is where they influence the initial decision on just what sort of equipment will best satisfy the operational requirement being considered. In order to give an overview of the role and application of these techniques in defence procurement, it will be helpful to outline the procedures through which a requirement for a defence system passes before a firm decision to order goes ahead. The important part played by considerations of reliability in the selection of equipment and contractor is then discussed, and, lest the reader should be lured into believing that all the problems of defence procurement can now be expected to be under control, the expected drawbacks will be outlined as well.

A particularly useful feature of life cycle costing is that it calls for the construction of a profile of the rates at which money will have to be invested throughout the entire life of the equipment. This enables a long-term view to be taken by the financial managers as the project develops, and shows up the potential weakness of applying short-term expedients in times of crisis. Experience of applying these principles in the defence industry in the USA shows that 70% of the total cost of a new project may have been committed before the production stage is reached.

This process means that the case for funds being allocated at a certain rate has already been argued and agreed in advance. It therefore helps to keep the project nurtured and viable during its early life when it is probably at its most vulnerable to cancellation or, at least, starvation of funds. The down-side of this, however, is that there will be times of governmental stringency during which the idea of early cancellation may suddenly become very seductive because of the huge savings in expenditure that will apparently result.

Another great benefit of preparing a detailed spend profile in advance is that the principal cost drivers of the project can be clearly identified at an early stage, and so can be modified if necessary while the procurement process is still firmly under control. A development/spend profile can also become an invaluable tool later in the project's life when examining the case for making major design improvements or even changes of role. Such reassessment exercises may be necessary when considering a mid-life update. Sometimes they may be forced upon the operating service if signs appear of unforeseen structural failure, which could lead to the equipment being withdrawn from service earlier than expected unless extensive repair and rework operations are carried out.

Many aircraft, ships, and fighting vehicles undergo at least one planned mid-life update in which they are refurbished and fitted out with up-to-date weapon systems and sensors. This is an economically attractive alternative to developing a replacement system from scratch. Warships, in particular, will come into service with their planned life of major refits and update programmes already mapped out in front of them. However, it should be noted that stretching the life of equipment beyond the point at which it had been planned to end may invalidate those earlier calculations in which the value of the planned life had been an important factor. This is where it is wise to have carried out sensitivity analyses.

Life cycle costing has come into its own as a discipline in times in which defence projects are highly expensive and very slow moving, and the numbers of units procured may be relatively small. Certainly, in the UK procurement numbers have been measurable in handfuls and penny packets compared with the speed and volatility of corresponding projects before the Second World War. For instance, the UK, which built over 50 000 military aircraft during World War 1, managed to deliver scarcely 80 in the year 1983, and here is a further reason for life cycle costing not having been relevant before. During the 1930s the Royal Air Force, for instance, had as many as 75 different aircraft designs in its inventory, and of these only eight were produced in serious quantities. Contract numbers varied alarmingly from year to year. Orders for the basic Tiger Moth training aircraft ranged from as few as two per year during times which were particularly lean to 2000 per year once the war was under way. Any attempts to apply the disciplines of life cycle costing to the business of defence procurement in those conditions could never have been successful.

5.3 The defence procurement process

The process by which the UK Ministry of Defence procures a new weapon system today follows a route each step of which is taken with caution and deliberation, with many pauses for consolidation and reconsideration. The ultimate destination has to be correct, and seen to be correct for a very long time. The process starts very early, and it may come as a surprise to the reader to learn, for instance, that the existence of a new aircraft being developed by the Ministry of Defence does not necessarily arise from a process that began with the intention of procuring a particular type of aircraft, or even an aircraft at all. Instead it will itself be the product of an earlier process of discussion that began with examining all the possible solutions to a particular problem, and by taking everything into consideration, life cycle costs included, which resulted in the decision that the best solution would take the form of an aircraft rather than a missile system or a ship.

5.3.1 Gestation process

The gestation process of a new equipment project consists of a number of phases which together are called the procurement cycle. Most major projects, and certainly those with high estimated development and production costs, will have to go through every stage of this, but some may overlap or be omitted altogether if this is seen to be likely to be more efficient than rigid adherence to all the rules. The essential principle is that the procurement process will go ahead in a step-by-step manner, in order that progress may be reviewed at clearly defined junctures or when a need is recognised for significant changes to be made.

The need to procure a new suite of defence equipment may arise from a number of different sources. These may include any of the following: new intelligence that the potential enemy is developing a weapon that cannot be countered adequately by anything in service at present; the recognition that current equipment is coming to the end of its life; a change in national or NATO defence policy; a proposal from the defence research programme or from the defence industry offering a new means of meeting an existing need or providing a new capability.

Ideas arising from these factors are expanded by exchanges of views between the staffs of the arms of the service that will be the likely customer, the defence scientific staff, the MoD procurement executive and its research establishments such as DRA Farnborough and RSRE Malvern, the views of the potential user usually being dominant. Then, following further analyses and mathematical modelling exercises the shape of the most suitable equipment finally emerges.

The development and growth of defence equipments usually follows a process of evolution rather than revolution, so usually aircraft do get replaced by aircraft as the defence manufacturers continually update their product. However, sometimes totally new systems emerge, such as so-called smart weapons and stealth aircraft, while old concepts like hovercraft, seaplanes and barrage balloons simply fade away.

The outcome of this initial process of concept formulation will be a decision that leads to the specification in broad terms of the functions and desired performance of the new aircraft or system. It is presented in the form of what is known as a Staff Target. The next step is to identify and bridge the gap between what is ideal and what is practicable. This is taken in the course of the next phase of the project development, which is the Feasibility Study.

5.3.2 The Feasibility Study

The Feasibility Study is the stage at which the defence manufacturer becomes involved, as it is the subject of a competitive contract for which tenders are invited from industry. Sometimes the Feasibility Study contract is awarded to two or more companies so that a selection of results may be obtained. Studies are relatively cheap to commission compared with production contracts, and at this early stage the more quantified information that is made available the more wisely the decisions on where to go next can be debated. Failure to identify inherent deficiencies in the project concept at this stage can have far-reaching adverse effects on later phases of the procurement cycle.

The Feasibility Study goes beyond the task of examining the practicability of the desired system. It also calls for preliminary estimates to be made of development costs and production costs, together with a forecast of the timescale of the realisation of the project. Note that the cost element is already becoming involved as a part of the whole study.

The results of the Feasibility Study provide the Ministry of Defence customers with information with which they can determine where to strike the balance between what they want and what they may reasonably expect to get. The practice of letting the Feasibility Study contract out to more than one company also gives a lead to indicate which defence manufacturer may be best placed to undertake the next and perhaps subsequent stages of the process.

The results of the study now enable the military customer to start putting flesh on the bones of the bare Staff Target. It is redrafted and issued in an enhanced form under the title of a Staff Requirement. Here the basic specifications of the equipment have become more firmly set, and the requirements to be satisfied by its other properties are set out. As well as the central parameters called for, such as range, performance, and physical dimensions, the Staff Requirement will specify items such as compatibility with NATO standards for servicing and armaments, together with values for the all-important performance characteristics of reliability and maintainability.

In the past, i.e. up to the 1970s, the values of these parameters only became known as experience was gained of the performance of the equipment in service. If they were unsatisfactory, as was often the case, the only way to salvage the situation was to commission a ruinously expensive programme of retroactive modification, which sometimes meant that the manufacturer was paid twice, the first time for supplying the equipment, and the second time for getting it to work. The lesson that support costs form the greater part of the total cost of ownership was first learned at a time when they were reaching runaway proportions.

The outcome of the experience has been a decision to include quantitative values for reliability and maintainability in the equipment specification, and to make the manufacturer contractually responsible for meeting them. As might be expected, it has proved easier to ask for this service than to obtain it. Nevertheless, manufacturers have been faced with no option other than to accede to these requirements if they wish to stay in business, although it has taken some twenty years for them to develop the necessary skills.

Then, hardly have manufacturers come to grips with contractual demands that make them accountable for the reliability and maintainability of the weapon system, than they find themselves faced with the responsibility for its performance as well. This development is discussed below.

5.3.3 Project Definition

Formal official endorsement of the Staff Requirement is followed by an invitation to the defence manufacturers to submit tenders for the task of taking the project into its next stage. This is Project Definition. Now the successful company can get down to the real hard work. They have to sort out the remaining areas of technical uncertainty, draw up a detailed and phased development cost plan and timescale for the project, which will be used as a basis for monitoring their progress, and make an estimate of the unit production cost. This figure depends, of course, very much on how many units are going to be ordered, and is likely to change as the project progresses. It is essential that the project is adequately defined in this phase. Insufficient definition will have an adverse effect on the technical programme, costs and duration of full development, and could have serious repercussions on the production process plus the date on which the equipment is introduced into service.

In really extensive projects, Project Definition (PD) may run to two stages, PD 1 and PD 2, with the whole task being reconsidered at the end of PD 1. Getting the whole procurement process right at this stage is considered as being so important to the success of the project as a whole that it is recognised as acceptable that as much as 25% of the total development cost may be invested in it.

5.3.4 Full Development

When, and only when, Project Definition has been satisfactorily completed, the equipment goes into the stage known as Full Development. This is normally, but not inevitably, undertaken by the project definition contractor on the basis of the technical, time, cost and management plans prepared during the Project Definition phase. The full development of a project involves all the engineering processes, trials and tests necessary to establish that the final design is capable of satisfying the Staff Requirement in all respects, and demonstrating that it is capable of being produced economically. This process will include the manufacture of models and prototypes, together with preparation of adequate logistic support backing in the form of handbooks, documentation, spares support packages, test equipment and facilities, and training aids.

Now responsibility for the project is passed on from the armed services departments to the Ministry of Defence Procurement Executive. They administer the trials and demonstrations necessary to satisfy the user service that the new equipment really can meet all the standards and objectives prescribed in the Staff Requirement. Often user trials are also arranged to validate the performance of the equipment in a service environment. Only when these processes are complete, and development has proceeded to the point where there is sufficient confidence that a standard acceptable to the user can be achieved, does the equipment go into production.

In practice it is seldom possible to effect a clear-cut transition from development to production. In many cases the ordering of long-lead-time materials and tooling for production, together with initial spares provisioning, must be undertaken in parallel with development if the required in-service date is to be met. This overlap entails risks which must be quantified and whose implications must be assessed. In order to minimise these risks, commitments to early production are usually entered into on a stage-by-stage release basis.

What has been described so far is the structure of the main stages through which the procurement activity has to pass. In practice, not every project has to pass through them all. The essential point is that no step is taken from one stage to the next without the collective approval of the relevant authorities within the hierarchy of the Ministry of Defence structure.

The time taken for the whole process, from concept to first delivery, can easily be as long as ten years. For example, the Tornado programme started in 1969, and the first aircraft began to come into service with the RAF in 1980. The rate of delivery then became established at about four aircraft per month, so, with the size of the initial RAF requirement being 385 airframes new aircraft were still expected to be arriving up to nine years after that. Needless to say, the aircraft the service would like to be receiving at that point would not be to the specification that had been drawn up twenty years earlier, and therefore the procurement cycle would have to have undergone a process of continuing revision.

The procurement cycle has to strike a compromise between the conflicting requirements of speedy delivery of the required new system and cautious deliberation over every step that is taken. Inevitably it takes a long time, and the longer it takes, the more essential it is that its progress is monitored by a method of financial discipline that can keep it under control.

5.4 Funding defence expenditure

When trying to rationalise the process of allocating and analysing the costs of a defence project it is important to remember that national defence is not a profit-making activity. A commercial enterprise can present accounts of income and expenditure which its shareholders can use to measure how well it is performing. The Ministry of Defence produces only expenditure. The price of investing in defence can be identified, albeit with difficulty, but the price of doing without it may range from absolutely nothing if it never has to be put to the test, to something incalculable if its non-existence results in defeat and destruction.

Nevertheless, defence costs have to be brought into the arena of public debate, and some realistic yardstick has to exist if they are to be seen to be reasonable but not extravagant. In the absence of anything better, the usual criterion for approving projected defence spending in the future is what was thought to have been acceptable in the past. So the amount allocated to defence spending every year is broadly based on how much was spent the year before. The government of the day wants to get away with spending as little as possible on defence; the defence chiefs on the other hand are unwilling to allow their forces to be diminished. In conditions in which the potential and strategic balance remains static year upon year, the main thrust of the exchequer is directed towards supporting the defence posture while minimising the amount to be spent on it. In other words, if defence spending cannot be reduced then the drive is towards making it more efficient. That is why the application of techniques such as life cycle costing is beginning to receive such active encouragement.

However, there is still a long way to go before life cycle costing can become properly established. One problem is that while the successful application of these techniques is vitally dependent on the availability of the costs of procurement and support of defence equipments, the cash elements of many of these remain very hard to ascertain. Another is that the present method of allocation of defence funding makes it very difficult to guarantee the stability of the long-term cash-flow profile which forms the very core to the process.

It might be assumed that in an environment that tries to be as cost-conscious as the Ministry of Defence, it should be a straightforward matter to identify all the components of the costs incurred in the procurement process. Unfortunately for the would-be researcher, auditor, or reformer, this is far from the case. Defence contractors seem to be very slow in presenting their accounts, being several years late in many cases, and tend to present them in aggregations such that the process of allocating elements of an account to specific activities is difficult, if not impossible. Recent developments in this area are, however, encouraging. Since 1989, the defence industry has been required to price all spare parts supplied to the Ministry of Defence.

The government allocates money for defence spending on a year-by-year basis. However, as we have seen, the procurement cycle of most, if not all, defence projects is more likely to be one of ten years or more rather than just one year. This has been acknowledged for a long time, and those departments within the Ministry of Defence which handle costs and budgets, as distinct from operations and policy, when presenting their annual submissions for their share of the defence budget are permitted to bid for funds up to ten years in advance. Every year they present their long-term costings so that the processes of specifications, design, development, and delivery of the defence equipments in their domain do not have to stop and restart. This is where the practice of allocating funds at as early a stage in the life of the project helps to ensure its survival when times are hard.

As well as regularly submitting their long-term costings, the sponsoring departments have to present their projected spending figures for endorsement by higher authorities on certain other specified occasions in the procurement cycle. These include the initiation of the project, the beginning of the next phase, the likelihood of predefined tolerance on cost or time being overrun, and revisions of the requirement itself.

Unfortunately, the principle of apparently approving long term costing for ten years in advance falls short of the ideal of being able to match them with a projected life cycle cost profile. Both governments and their financial circumstances are subject to change from year to year, and it is far from unknown for a demand to be made that, for instance, there is to be an instant 10% cut made in defence expenditure right across the board.

Those offices which are concerned with an emerging project are, of course, no more immune from strictures of this nature than are any other departments, and they therefore periodically find themselves having to juggle funds from one year to the next and back again in an attempt to keep the project afloat. Unfortunately, although, as has been pointed out, the various stages of the procurement cycle through which the project passes are not rigidly compartmentalised, the same is not so for the budgets that they control. This can mean that an economy applied by one office can reappear later as an additional cost to be borne by another. This is especially prone to happen when a cut in investment in reliability and maintainability made at the procurement stage results in an increased cost of maintenance to be borne by the operating service.

Another problem in this context is that, under the present rules, the residue of an underspend in one particular year may not be carried forward to the next. This, of course, encourages profligacy at the end of every financial year, when departments hastily spend their full allowances in the knowledge that they will lose them if they do not, and militates strongly against the very principles on which life cycle costing is based.

5.5 Forecasting the life costs of a project

The main areas of costing that are subjected to analysis and forecasts early in the life of a new project may be considered under the headings of first costs and operating costs. These are not, of course, independent. The very basis of life cycle costing is that it is essential to consider operating costs when first costs are being optimised. One broad way of differentiating between the two is to suggest that first costs are more closely associated with the manufacturer while operating costs fall to the user.

5.5.1   First costs

Manufacturers continue to develop and refine their own methods for the estimation of their contribution to the cost equation. These are usually based on some mixture of parametric studies, i.e. cost = some function of size, speed, weight, labour rate, batch size, etc. and near neighbour comparisons, i.e. how much did it cost to build a similar one? Cost estimation processes corresponding to these are used when a company is engaged in the preliminary stages of the procurement cycle, such as Feasibility Studies and Project Definition.

5.5.2   Operating costs

Support costs must be assessed by the potential user, and they therefore will be particularly concerned with those aspects of the specification and design processes which are going to have an effect on these. These comprise, in ascending order of difficulty of identification and quantification: the costs of providing static support requirements such as docks, airfields, hardened shelters and ground equipment; the costs of crew numbers and their training; costs of consumables such as fuel, oil, and tyres; costs of scheduled maintenance and support; and, most difficult of all, the costs associated with unscheduled maintenance, i.e. unserviceability.

The enormous significance of these support costs, in that they comprise the major part of the total life cost, has already been emphasised, and therefore it is essential that they can be forecast and methods developed for keeping them under control, even before the equipment has been designed. This is the area in which the superiority of the maintenance policies and practices of aerospace equipment compared with those applicable in sea and land settings shows its worth.

The task is to find out everything that can be learned about the shortcomings of the equipment that is in service today, and to apply those lessons when writing the specifications for the equipment that it is intended to bring into use tomorrow. Ideally this information should be readily available for all equipment, regardless of the medium in which it operates; the reality is that, so far at least, it is only in the air world that the necessary data are recorded and maintained in an accessible form.

Historically, this came about because when flying machines first appeared in this skies they were immediately seen to be more threatening by far to life and property than any other self-propelled vehicle with which society had become familiar. Legislation was therefore demanded to ensure that of all the considerations governing their design and operation, the principal consideration should be safety.

The result was that national standards were set up. These standards were to regulate the design of aircraft, the materials from which they could be constructed, and the qualifications of individuals permitted to fly them, build them, repair them, and service them. The most important part of these standards in this context was the specifying of the formal records that must be kept in order to verify that all the rules were being strictly enforced. In simple terms, all work on aircraft has to be written down and signed for.

These records continue to be maintained today for all current military aircraft and weapon systems in the UK, and indeed in most of the developed world. For many years data abstracted from them have been used to build up a comprehensive reliability database, which is available for ready access by defence and design authorities. This reliability information is in constant use for refining the operating and support practices for aircraft in service now, and planning for those of the future. Operators of defence equipments on land and sea have been quick to recognise the benefits of having such databases available, and the formal intention of applying through-life costing techniques to all aspects of defence procurement is helping to speed their introduction.

5.5.3 Serviceability

The characteristics that may be used to quantify the serviceability of aircraft and other equipment are reliability, availability, and maintainability. The management of reliability has now reached a stage of development, in aircraft at any rate, such that the services know what minimum values will be required to ensure the degree of mission-effectiveness they need, and can specify it in the confidence that the manufacturer will be able to supply it. Maintainability, which is a measure of how easy an aircraft is to service, is built in as a design feature, taking account of the requirements for rapid replenishment and easy access for servicing and repair. For example, about 45% of the surface area of the Tornado consists of removable panels giving access to its internal systems, 90% of the units thus accessible are located at normal working height, and no formal maintenance other than flight servicing is called for at intervals of less than 400 flying hours. In less developed aircraft in the past, no serious consideration whatever was given to factors such as these.

Reliability and maintainability combine to support the most important parameter of all, availability. This is a measure of how likely the aircraft is to be ready for service when called for. Its application can be illustrated by means of a very simple example. If the forecast availability of the fleet is 50% and the operational scenario on which the system procurement has been based calls for 30 aircraft, then the number to be ordered will have to be 60. However, if the availability is increased to 60% then the task can be met by 50, a saving of 10. Therefore, if the user service can specify values for reliability and maintainability which they know from experience to be reasonable and capable of achievement, and which the defence contractor can be expected to deliver, then the user service can limit the order to a minimum size in the confidence that they are risking no loss of effectiveness. The importance to the successful implementation of life cycle costing techniques in defence procurement of understanding reliability and maintainability and how they relate to availability, of having access to historical data on which to base quantified specifications that a defence contractor can confidently be expected to meet, and of maintaining and analysing fresh data as they arise cannot be overstressed.

The significance attached to support costs may be recognised from the emphasis placed on them in the publicity material produced by the manufacturers of defence equipment. The principal selling point of a military aircraft is no longer its impressive performance in flight and weapon delivery, instead it is how well its support costs undercut those of its competitors. While the UK Ministry of Defence is very cautious about publicising any features of aircraft performance such as this, and would certainly never permit operating costs to be divulged, no such qualms affect the US Department of Defense. Service targets for maximum costs per flying hour are freely quoted in press reports on defence debates and transactions, alongside figures relating to manufacturers' fixed-price guarantees for various rates of production and fleet sizes. Also included are their forecasts of costs per flying hour and costs per engine running hour in current dollar values, and their predictions for 25-year life cycle support costs. Such openness would be very welcome in the UK, if only to help confirm that defence contracts are awarded on grounds of performance alone.

5.6 Defence contracts

A significant factor in managing the costs of defence materials is the way in which the contractors charge for their services. For many years the flexibility that is bound to exist in the processes of specification, procurement, and especially development, plus the potentially disruptive effect that this might have on the contractor's own financial management, was recognised in the way they were allowed to charge for their services. Contracts were awarded on a basis of 'cost plus', which meant that the contractor charged for the work done, plus a fixed percentage as profit. This could easily mean that if, for instance, a protracted programme of research failed to reach a satisfactory conclusion, or if the customers kept on changing their mind about the specification, even though no satisfactory end product was delivered the contractor continued to run a profitable business.

Cost-plus contracts could have unfortunate consequences, the most

notorious of which, in the UK, was exemplified by the affair of the airborne
early warning (AEW) version of the Nimrod maritime patrol aircraft. When
development of this aircraft started in 1977, costs were estimated to be £300
million. By 1986 no less a sum than £1 100 million had been spent, and the
aircraft still failed to meet its performance requirements. At that point the
project was cancelled, with much mutual recrimination between the Ministry
of Defence and the contractors on the questions of what was the aircraft
really supposed to be capable of, and who was responsible for ensuring that
it could do it.

One of the reasons for the failure of the AEW Nimrod was the fact that
the Ministry of Defence held one company, British Aerospace, responsible
for developing the aircraft, and another, GEC Marconi, for the radar systems.
Nobody had clear responsibility for the overlapping areas where problems
might occur, giving rise to a conflict of interests, and nobody had laid down
a clear measure of precisely what system performance would satisfy the
specification.

This fiasco led to a number of fundamental changes being introduced into
the Ministry of Defence contracting procedure. One was that the system of
cost plus was abandoned, and all future contracts would be fixed price.
At first sight this appears to be no more than reasonable, but it brings further
potential problems in its train. Now a defence manufacturer preparing a bid
for a fixed-price contract has to make a forecast of their own likely expenditure
for many years in advance, and scale the price up in such a way as to leave
them still safely in profit at the end of the day. On its part, the Ministry of
Defence may get a bargain, but it also has to face up to the possibility that
serious cost overruns could bankrupt its contractor, leaving the Ministy of
Defence with the unhappy choice between having to incur further expense
by having to bail its contractor out, or doing without its weapon system
altogether.

Another development is that the contractor can now be required to mount
a satisfactory demonstration of the weapon system against stated performance
objectives. This has come about following the delivery in past years of a
number of systems that have failed to work as specified, putting the user
service in the position of having to complete the development process itself,
sometimes by raising further cost-plus contracts on the original supplier.
However, specifying weapon system performance in such unambiguous detail
that it can form part of a rigidly enforceable contract is no easy task. In one
current system procurement, the job has been contracted out to a defence
company.

Current indications are that defence companies on both sides of the Atlantic
are becoming increasingly wary of bidding for prime contracts, knowing the
risks they may entail, while the defence ministries are learning that their move
to unload all the risks on to the industries may not have been so sensible
after all. A defence industry that has been trimmed of all its fat no longer

has a surge capability with which to expand in times on rising international tension. It is no wonder that the Boeing Aircraft Company once suggested 'Tell us the targets, and we'll deliver the bombs on a cost-plus basis!'

It is possible that the switch from cost-plus contracting to fixed-price contracting is having a detrimental effect on the defence industry more severe than the imposition of belt-tightening measures alone, and the search is on for a new structure of contracts that will include incentives for good performance. All this activity can only make life cycle costing more difficult, even though in the long term it will be more effective.

5.7 Summary

The procurement and operation of defence equipment is a very expensive business. For many years all efforts to economise were directed towards minimising only the first cost of the equipment. This was an appropriate target when in-service lives were short, but now that some equipment is expected to last for 25 years or more it has become evident that the first cost represents only a relatively small proportion of the cost of total life ownership, and is far outweighed by the support costs. It is now recognised that financial control will be best achieved by the application of life cycle costing techniques.

For life cycle cost management to work it needs data to be available on the costs of aquisition and support, plus the guarantee of a predictable cash-flow profile over many years. Given these, LCC can be readily superimposed on the procedures for defence procurement currently in use, and its benefits made available to government and taxpayer alike.

The process of the efficient application of life cycle costing to defence equipments is a journey rather than a destination. No staff officers can forecast what is going to happen to the equipment in 25 years time or commit their successors to scrapping it at the age at which the minimum cost predictions would have been accurate. It will take a strong organisational discipline to ensure that this deliberate policy of 'spend now to save later' is not allowed to degenerate into one of 'spend now and spend later as well'. The signs are encouraging that this discipline is being imposed.

6 The quality approach to design and life cycle costing in the health service

J.F. McGEORGE

6.1 Introduction

It is widely recognised that the quality of design is crucial to the success of the construction or production process. Fairly minor changes in design (e.g. constructibility reviews) can have major effects on the cost and efficiency of production or construction as well as on the usefulness and marketability of the product. The design paradigm can permeate whole companies (Philips is usually quoted as an example) or whole countries (Italy is well known for quality of design in many fields). It should be noted that the word 'design' is used here in its widest sense, i.e. including everything that is done by design rather than by accident. In this context, the definition would include:

(a) Determination of the need to be filled.
(b) Conceptualisation of a solution.
(c) Embodiment and detail design.
(d) Consideration of the effect and usefulness of the product.

6.2 Design and the life cycle

In recent years the question of life cycle costs on constructed projects has come increasingly to the fore, mainly due to changes in the relative costs of the inputs. In particular, the cost of materials as a proportion of the costs has declined, while labour, finance and energy costs, among others, have risen.

The design approach is, of course, critical in determining the construction costs as well as life cycle costs. It is not the intention here to discuss in detail the means by which life cycle costs may be calculated and reduced—that has been competently done elsewhere—but rather to discuss the difficulties of implementation, and suggest approaches for improving implementation. The following points will be developed:

(a) Designing for life cycle costs is a matter of resource allocation and deciding on priorities.
(b) Many of the difficulties of implementation are caused by problems with the system which governs the decision-making process. This is a

major challenge, as changing systems requires radical changes in attitudes and bold decisions at the political level.

6.3 Designing for construction costs

Historically, much design has aimed at minimising construction costs, as expressed in the phrase 'An engineer is someone who can do for £1 what any fool can do for £2'. This frequently meant minimising material usage, and much effort was invested in sophisticated analysis procedures, which enabled structures to be designed to ever lower safety factors, thus reducing the material usage. The thinking here was of course partly governed by the idea that a safety factor was an allowance for ignorance—thus, a more detailed analysis which reduces the ignorance can lower the safety factor required. In this context, it has often been noted that many older bridges are over-designed, even though they were designed for lower traffic loads than exist today. A case was reported of one 75-year-old steel bridge which had suffered extensive corrosion but still retained an adequate margin of safety (Kuesel, 1990). This over-design frequently produces long-lasting or low-maintenance bridges, a facet much appreciated by engineers responsible for their maintenance. 'No one knows the true value of long-lasting bridge construction better than our engineers who maintain these structures' (Hahn, 1990).

Of course, it has long been recognised that the goal of lowest construction cost is often not well served by minimising material costs, as ease of construction (or constructibility) has a major influence on the total cost. The critical influence of the design on construction costs was pointed out in a paper entitled 'Designing to reduce construction costs' by Paulson (1976), in which he expounded the level-of-influence concept. This states that ability to influence cost decreases continually as the project progresses, from 100% at project sanction, to typically 20% or less by the time construction starts. The point of Paulson's argument is that the greater part of management effort to control costs is applied to the construction phase, where its potential effectiveness is very limited.

McGeorge (1988) extended this thinking, and developed the 'pyramid model' to explore the influence of design quality upon construction and other life cycle costs. The progress of a project from the initial idea to engineering reality is depicted by a process as shown in Figure 6.1. The term 'embodiment design' in Figure 6.1 is the one suggested by the British Engineering Council for that phase of the work which establishes the ways in which the conceptual design will be realised, and 'lays the foundation for good detail design through a structured development of the concept' (Engineering Council and Design Council, 1986). There are other aspects which need to be added to Figure 6.1 for a life cycle analysis; these include commissioning, operation, maintenance

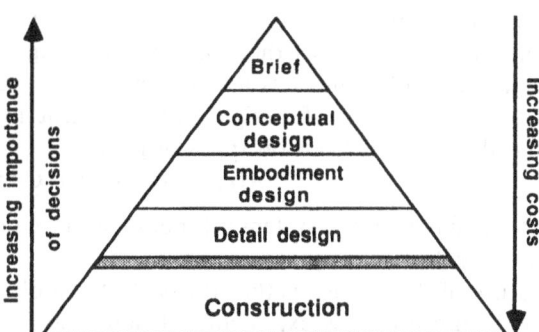

Figure 6.1 Design in the project process (McGeorge, 1988).

and decommissioning. These have been omitted for the moment in order to focus on construction costs, because they vary greatly in importance depending on the nature of the project. It now appears, for instance, that the problem of decommissioning was a major oversight in the design of the first nuclear power stations, thus presenting engineers with a major problem.

Considering initially the stages shown in Figure 6.1, some important features can be noted. Firstly, information flows both up and down the chain, to fuel the process. 'Design is an iterative process with each iteration aimed at increasing the level of information in order to improve the decision making. Coordinating the collection, processing, storage and transmission of information is essential for effective design. Existing information flows should be analysed to identify bottlenecks and remove them' (Engineering Council and Design Council, 1986).

In the light of this description, the gulf between design and construction, across which information flows only with difficulty, is an obvious anomaly. This gulf, resulting from the traditional separation of the design and construction phases, is a consequence of the structure of the construction industry. Researchers of the Construction Industry Institute (CII) in Texas state that this separation 'opposes many project objectives. It is neither advisable nor necessary' (Krizan, 1986). The same point is made by Taylor (1985), who notes that the interfaces in the construction process may serve individual purposes, but, in totality, impede achievement.

Much attention has been focussed, in recent years, on ways of reducing or overcoming this separation. Various approaches have been used to improve management and cost control and involve the contractor in the design stages (Heinen, 1985; Kirschenman, 1986). These changes usually entail alternative contractual arrangements, and it is partly in response to this trend that the New Engineering Contract has been developed in the United Kingdom. The following quotes from the 'Need for and features of the NEC' illustrate the point.

'The industry has developed, and employers are frequently using a much wider range of contract strategies than before, including management contracts, design and build contracts and target cost contracts.'

'The traditional separation of design from construction has been questioned, and discarded for certain types of project.'

(*New Engineering Contract*, 1991)

It is apparent that change in the system is fundamental to improved design quality and cost control.

The second feature of the process is that the cost of completing each stage increases rapidly, in more or less exponential fashion, as indicated by the cost pyramid. The cost of design is generally considered to be roughly 2–10% of the total costs (Institution of Civil Engineers, 1985), and the implications of this are interesting. Since design costs represent only a small proportion of total costs, it becomes worthwhile to increase design effort significantly in order to achieve comparatively small reductions in construction cost. A simple example, using a design cost of 5%, is shown in Table 6.1. In this case it has been assumed that a 50% increase in design input yields a 10% saving in construction cost. The net result is a 7% reduction in total cost, or a saving of almost three times the cost of the extra design work. It should be questioned, of course, whether such figures are realistic, and the evidence available indicates that they are in fact under- rather than over-stated. Research carried out on constructibility, for example, indicates that constructibility and value-engineering reviews typically yield construction-cost savings of 10–20 times the cost of the extra design input (Business Round Table, 1982). Even these figures understate the true potential for improvements, as constructibility exercises, by definition, aim at reducing only construction costs, and the picture is dramatically improved by including operating and maintenance costs, which frequently exceed construction cost. To take one example, figures on hospitals indicate that operating costs exceed the capital cost of building a hospital within only 2–3 years of operation (National Building Research Institute, 1985a).

Naturally, improvements cannot continue to be made indefinitely. At some point an increase in design effort will yield an insignificant saving on construction, and the total cost will be higher. Thus, plotting design input against

Table 6.1 Effect of extra design input

	Costs (£ thousand)	
Item	Original design	Revised design
Design cost	50	75 (+50%)
Construction cost	950	855 (−10%)
Total cost	1000	930
Overall saving		70 (7%)

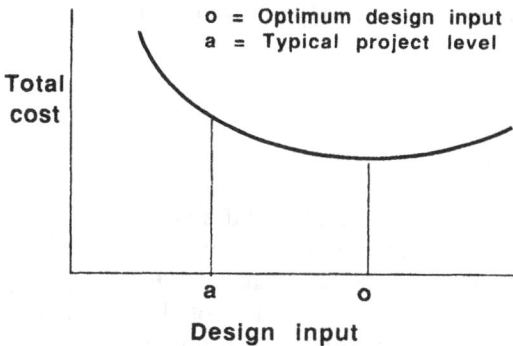

Figure 6.2 Optimum design input (McGeorge, 1988).

total cost yields the curve shown in Figure 6.2. The rational policy then would be to aim for the optimum design effort 'o', but the implications of the foregoing discussion are that most engineering projects fall well to the left at some point 'a'. The success of constructibility programs is a result of recognising and acting upon this simple and obvious fact.

A further implication, and one well recognised by most workers in constructibility, is that the largest gains can be made early in the process or high up on the scale of 'Importance of Decisions' (Construction Industry Institute, 1986). The author has at present no figures on the cost of conceptual design as a percentage of total design cost, but a relationship similar to that between design and construction may be reasonably inferred.

A caution is needed here, as putting extra resources into the design will not automatically improve construction or life cycle costs. It is possible that making extra money available for design will simply increase the cost and the time taken, without any concomitant benefits. There is undoubtedly truth in this, as was recognised years ago by Parkinson (1960) when he formulated his famous law 'Expenditure rises to meet income', but there are two errors which have grown from this line of thinking. The first is the idea of a corollary to the law, namely that if income (i.e. resources for the design) is reduced, then expenditure will also fall, without any adverse effects on the quality of the product. The second error is to overlook the benefits that can be derived from proper attention to the critical early stages, at the top of the pyramid, for this is where the extra effort needs to be focussed.

Constructibility reviews have been successful because they focussed on design deficiencies — in this case the lack of attention to the construction process. In the same way there are frequently deficiencies in establishing needs, and thus the requirements of the design. In discussions held with officials from the Department of Works (responsible, amongst other things, for construction of hospitals in South Africa) it was observed that a frequent cause of overspending was lack of understanding of needs by designers. In order to function effectively, designers need a clear, positive directive, i.e. an excellent brief.

6.4 Design quality

These considerations lead to the idea of design quality—a good design will be effective (i.e. serve the purpose for which it was intended) and constructible with the best possible economy and safety. Thus the quality of design is a major determinant of construction and whole life costs, and to improve life cycle costs it is necessary to improve the quality of design. That this is a problem there need be no doubt. Vlatas (1986) points out that design deficiencies are a major cause of contract disputes and changes during construction, and notes several causes for the increase in design deficiencies. From the previous discussion, it follows that quality is a management problem, and primarily one for top management. For it is at the top, in the early stages of agreeing on the terms of reference and scope of work and determining a conceptual design, that the really important decisions are made. Parkinson (1967), analysing executive remuneration, concludes that, at the top level of decision-making, the policy decisions become so important that the cost of making them is almost irrelevant. It is not for nothing that people like Michael Edwardes are paid six-figure sums for less than six months in the top position, as happened at Dunlop. The Dunlop shareholders contemplating the returns should feel it is money well spent (*Financial Times*, 1985).

Ideally this process should extend even further back, to the determination of the market need, as described in the Engineering Council's model. 'Should the English Channel be crossed by a bridge or a tunnel?' is a question of conceptual design (with many implications besides mere economics). The market-need stage addresses such questions as 'Should the crossing cater for road, rail or both?' and 'Should there be a crossing at all, or should reliance be placed on the ferry service?' These are most important questions, but such decisions, even on projects much smaller than the channel crossing, are usually political, and are thus largely out of the control of engineers, apart from those few willing to brave the political arena.

It can thus be seen that, even when considering only first construction cost, the following points emerge:

(a) The quality of the design is critical in determining costs.
(b) It is the early, conceptual decisions, highest up the scale of importance of decisions, which require the most attention.
(c) Designs are seldom optimised in the way indicated above, because this is discouraged both by prevailing mind-sets and by the system. There is a lack of appreciation among clients and owners of the value of

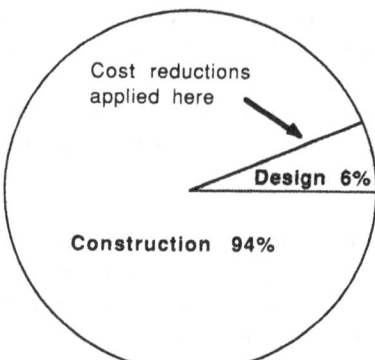

Figure 6.3 Misdirected cost reductions.

professional services, including design services. This is viewed as largely unproductive expenditure, and is therefore skimped as much as possible. The folly of this is easily illustrated by a pie chart (Figure 6.3). Using a typical design proportion of 6%, it might be imagined that an owner would consider means of reducing the size of the large construction slice, as described in Table 6.1. The usual practice, however, has been to go the other way and attempt to reduce the 6% to some lower figure. This tendency extends right down to the private house owner, and the author has observed cases where an owner spends a large sum on the right piece of land and budgets generously for quality construction and finishes, but baulks at paying architect's fees to get a quality design. Invariably those who do pay the price end up with a more satisfactory result, and frequently, reduced costs into the bargain.

In addressing the above difficulties at construction cost level, the following systemic factors can be noted:

(a) The competitive bidding system for construction contracts, which separates design from construction, creates difficulties in integrating design and construction for the best result. Many papers and articles have been published in the search for a better system (e.g. Nicholson, 1991).

(b) The system of payment for design (usually a percentage of construction cost) has the effect of discouraging the designer from investing the effort to make improvements. 'It has to be acknowledged that commercially oriented people find it a very strange notion that an engineer would work harder and spend more man-hours in honing his design, thereby reducing the construction costs, and the engineering fee! To work more, in order to receive less is an alien concept that no profession or business, other than consulting engineers, can subscribe to' (Shepherd, 1987).

(c) The method of selecting a consultant also poses difficulty, particularly for public-sector clients, where the possibility of corruption has to be visibly excluded. The effect of this has been discussed more fully elsewhere (McGeorge, 1988). Where the decision-makers are in the private sector, spending their own money, such considerations may not obtrude to the same extent.

In summary, the fact that the design phase is not managed to produce the minimum total or whole life cost is an inevitable consequence of the way the industry is structured. It is a consequence of the fact that design and construction are treated separately; the costs of the design are negotiated with the consultant in isolation from the costs of construction, and prior to the start of the design. The result is suboptimisation of the design phase.

6.5 Designing for life cycle costs

All the above considerations apply when extending the argument to encompass whole life costs. Again the design is crucial, and requires a clear understanding of the objectives and economics of the project. Because circumstances and economics change, to do this satisfactorily sometimes requires a designer to be blessed with the vision of a prophet! Prior to the mid-1970s the USA was in a position where building materials were comparatively expensive, while oil, and thus energy, was cheap. Many buildings were then constructed with minimal use of materials and little insulation, relying on heating and cooling to keep them comfortable. The energy crisis of the late 1970s changed all that and made many of those uninsulated buildings costly to run. In the UK, where energy has never been quite so cheap, there has long been an awareness of the importance of insulation, even in domestic dwellings.

The same holds true for Europe, and the recently completed civic centre in Marseille is reported to have cut operating costs by as much as 45%. This 'bioclimatic' design uses a mixture of ancient and modern techniques to reduce energy demand. A simple example is the use of 500 mm-thick concrete walls to improve insulation—definitely not a case of designing for minimum construction cost (Civil Engineering ASCE, 1990).

6.6 Life cycle costs of bridges

The example of bridges is an interesting one and contrasts well with hospitals (considered later), as many bridges, once constructed, are low-maintenance structures, and life cycle costs are frequently not considered. Although some steel bridges require frequent repainting, the real life-cost considerations occur when the bridge requires repair or replacement, and it is here that the problems arise. The repair or replacement of a bridge is never as simple as the original construction, because the connection has usually become a vital part of a route, and traffic accommodation must be considered. It is worth pondering the likely effect on accepted design philosophies of accurately estimating and factoring traffic congestion and disruption costs into the life cost of roads and bridges.

In an article in *Civil Engineering ASCE*, Bettigole (1990) lists some figures with alarming implications for the USA. Noting that the average life span of USA bridges is about 68 years, while the decks last only 35 years, he states the following:

'Too often, this fact of life has been ignored by bridge designers and owners. Few bridges have been constructed with provisions for future deck repair or replacement, and fewer still with any thought given to rerouting or maintaining traffic while the work is under way.

Because most owners are public agencies, the only criterion, with few exceptions, has been lowest first-cost—as opposed to optimum life-cycle cost. This has included *lowest-cost design* and construction method as well as lowest-cost materials and systems. The present dilapidated state of America's transportation infrastructure can be laid at the feet of this time-honored philosophy.'

(Bettigole, 1990; present author's emphasis)

It is clear that neglect of the life cycle of construction works has potentially serious consequences. Every engineer and transport economist is aware of the critical importance of adequate transport links to a modern economic system. The effect of bad planning and design practices is to impoverish the whole country. Bettigole (1990) goes on to note:

'During the 1990s 40% of the total highway bridge deck area in the US will become 35 years old, statistically ready for replacement. Since any construction disrupts the flow of traffic, working on 40% of our total bridge area within one decade will require elaborate staging to keep our surface transportation moving. The need for scheduling such staging will only complicate each construction project and delay the solution of the bridge deck problem.'   (Bettigole, 1990)

As was the case when considering only construction costs, the blame for poor designs that do not meet life cycle considerations can be laid at the door of the system. In the USA the system is that most bridges are financed, at least partially, with federal funds, and achieving the lowest 'first-cost' construction becomes top priority, since the federal government does not share contractually in maintaining the bridge after it is built (Hahn, 1990).

The problem is not confined to bridges, nor to the tendency to design for lowest first cost. An equally severe difficulty faces road networks, where the problem is securing sufficient funds to maintain the network once built. In many countries financial stringency has led to cutbacks on infrastructure maintenance because of a lack of appreciation of the life cost implications and the prevailing short-term orientation of most political systems. This is particularly critical with infrastructure because of the wealth-creating effects of an efficient infrastructure, and concern is now being felt beyond the confines of the engineering world. An American general-interest journal recently devoted its cover story to 'America's crumbling infrastructure—losing wealth through the cracks' (*The World & I*, 1991).

For construction of roads and bridges, maintenance and repair are usually the major long-term costs, as the indirect costs and benefits of congestion or improved access, leading to economic development, which may be much larger, are rather difficult to measure, and subject to controversy. For some types of construction, however, the cost of running the facility once built far outweighs the mere capital cost of construction, and failure to appreciate this can lead to huge misallocation or waste of resources. The best example of this is hospitals, where, in most cases, the operating costs exceed the construction costs within 2–3 years, and in the case of academic or teaching hospitals, a period even shorter than this.

6.7   Life cycle costs in hospitals

The question of hospitals and their relationship to the rest of the health-care system is a most enlightening study, which highlights many of the issues raised in the pyramid model. A study carried out in South Africa in 1975 to determine needs and norms for hospital and health-care provision noted the critical importance of the top of the decision pyramid, and particularly the 'market-need' stage. It quoted the findings of a similar 1972 American study:

'The better the analysis of health care needs, the more highly that the appropriate service will be provided. Our study showed a need for improvement in both respects. Projects were often conceived in a crisis situation rather than in an orderly fashion; little or no attention was given to analysing the specific health care needs before planning the services to be offered and the facility to be constructed.'

(Webb Committee of Enquiry, 1975)

On capital expenditure (including construction costs) the report noted that the capital cost of hospitals and related services is a relatively small proportion (approx. 10%) of the total cost of hospital services:

'Economies achieved in the provision of effective preventive medicine coupled with the judicious deployment of medical and nursing personnel far outweigh economies that may be achieved by limiting the costs of building.'

(Webb Committee of Enquiry, 1975)

This view was confirmed by the National Building Research Institute (1985), which noted that capital costs then constituted 8% of annual health expenditure. In the case of a single hospital, capital costs amount to about 6% of the total life cycle costs.

It was stated earlier that life cycle costing is a resource-allocation problem, and this is recognised in health services as much as anywhere else:

'A coordinated, well-planned health service is essential to ensure optimal utilisation of available resources.'

(Webb Committee of Enquiry, 1975)

6.7.1   Academic hospital costs

In view of its importance in the health-care system, as discussed later, and its very high operating-cost to capital-cost ratio, an academic hospital makes a most interesting study. Exact costs are difficult to obtain, due to the lack of cost-centre based budgeting in most public hospitals. For the present, figures obtained from a sample hospital enable some tentative, but important conclusions to be drawn.

(a) Equipment costs, or the so-called 'non-consumables' (which include everything from soup bowls and linen to X-ray equipment and heart–lung machines) are a major part of the capital cost of providing a hospital. The cost of equipping an academic hospital as a percentage of the cost of building construction can exceed 70%, i.e. it comes close to doubling the initial capital cost, usually thought of in terms of construction costs only. Failure to plan for this can lead to waste and misallocation of resources. For example, South Africa's well-known Groote Schuur hospital has recently occupied a spacious new main building, but has had to resort to public appeals for money to purchase certain key items of equipment, due to funding cuts. In life cycle terms this is very short-sighted planning, as the expensive facilities cannot be utilised without the necessary equipment, and such shortages will ultimately drive away the skilled people on whom the whole system depends, as discussed in section 6.8 below.

Unfortunately, Groote Schuur is not alone in this predicament; the funding crisis is severely affecting the entire health-care system. One other lesser-known hospital currently under construction appears unlikely even to open, as there are insufficient funds to equip and staff it.

For non-academic hospitals, with less specialised equipment, the equipment costs are lower. A preliminary study of a small community hospital indicates a figure of the order of 30% of construction cost.

(b) As indicated earlier, operating costs exceed capital costs within 2–3 years of construction. As might be expected, personnel costs are the major portion of the operating costs; a rough breakdown for an academic hospital is as follows:

Personnel	66%
Consumables	27%
Equipment replacements	2%
Other operating costs	5%

A further breakdown of the first two major categories above yields the following information:

Personnel costs		Consumables	
Nursing	35%	Medical, surgical and	
Professional	31%	radiological supplies	49%
Technical	9%	Pharmaceuticals	31%
Administration and		Food	6%
general support	25%	Other	14%

The above figures are extracted from five years of running costs and are thus representative of one academic hospital in South Africa. It should be pointed out that these figures will not be reproducible elsewhere, as South African academic hospitals form parts of much larger complexes offering health care at many levels, and it is not possible to separate out the academic component. The figures are given purely for the purpose of highlighting the major cost areas. It can readily be seen that efficient use of nursing and professional staff should be primary objectives of the operational design of hospitals. Control of the major consumable items is perhaps more in the hands of the medical staff than the designer, but it is an item large enough to be worthly of consideration.

6.8   Academic hospitals in the health-care system

An academic hospital is interesting as a life cycle case study not only because of the great importance of life cycle costing within the hospital itself, but also because of its position in the health-care system as a whole. Just as conceptual policy decisions are at the top of the project-cost pyramid, so academic medicine and academic hospitals are at the top of the health-care pyramid, and thus, by extension, of economic activity and productivity in the country as a whole (Figure 6.4). This type of hierarchical model is well known, and a similar diagram was produced in a study on community health centres (National Building Research Institute, 1985b).

Looking at the pyramid, it can be seen that there is an increasing use of resources going downwards, but the most expensive and most critical resources are at the top. Thus, academic medicine is relatively more expensive than other parts of the system, but it is a smaller component, and consumes a relatively small proportion of the budget. This is easy to hypothesise, but figures for the relative amounts are not easy to come by, particularly in South Africa, where the so-called 'academic hospitals' actually supply a large number of services strictly associated with other parts of the system.

Figure 6.4 Health system pyramid.

In order to confirm the pyramid model, the strictly academic components of the services would have to be separated out from the various other services. This is not possible at present, but a number of public hospitals are moving towards cost-centre based budgeting, and the exercise may be possible in the near future.

If it is accepted that the model is reasonable, some interesting conclusions follow. The purposes of academic hospitals may be summed up as follows:

(a) To supply trained staff to the lower levels of the pyramid.
(b) To undertake research in order to supply knowledge, new treatments and techniques to prevent illness or otherwise reduce costs.
(c) To take on difficult, interesting or unusual cases which require specialised resources, and offer possibilities for research and teaching.

These functions have important life cycle implications:

(i) If these functions are not adequately fulfilled, the effectiveness of the entire health-care system will be undermined and, in particular, cost-effectiveness will suffer. Compare, for example, the cost of administering a vaccine with that of treating a disease, or consider the cost of ineffective treatment or incorrect diagnosis given because of inadequately trained staff.
(ii) Academic medicine is costly, as is university training in any discipline. However, it is not a luxury; it is a cost society (and South Africa in particular) must meet if it is not to decline into a wholly Third-World society.
(iii) Although costly in per-unit or per-patient terms, academic hospitals are not costly in overall terms, due to the pyramid structure. In order for this to hold true, it is necessary to limit the size and number of academic hospitals. It is necessary to provide adequately for the lower levels of the pyramid, as clinics and day hospitals have been found to

have a marked effect upon the demands made on an academic hospitals, and render a service at a lower cost (Webb Committee of Enquiry, 1975). The multi-level care currently provided by academic hospital complexes is often felt by the hospitals themselves to be inappropriate, but it is unavoidable because of the lack of facilities at other levels. In order to promote effective use of resources, it is necessary to make proper use of the cheaper primary health care.

It is important to note that the effects of academic medicine are long-term, and thus that inadequate funding will have no immediately observable effect. There is great danger here, arising from what has been dubbed the NIMTOO (not in my term of office) syndrome, to which politicians are particularly prone. Most engineers will be aware that the same syndrome also affects infrastructural construction and maintenance.

6.9   Effect of the system on costs

Since the major component of operating costs in a hospital is staffing, attention should be focussed on ways to make the most effective use of this resource. When the question is examined, it is apparent that once again we have a systemic problem, and the system needs to be reconsidered if any improvement in resource allocation and utilisation is to be realised. There are at least two important ways in which the structuring of the health-care system affects demand for expensive hospital resources.

6.9.1   Bottom-up health care

The health-care system needs to be constructed from the bottom up, so that the cheaper, lower levels of primary care perform a filtering function to reduce the load on the expensive tertiary levels. Figure 6.5 has been drawn from a model put forward by the Department of Health and Welfare (1981). Despite its similarity to Figure 6.4, Figure 6.5 is a bottom-up model, emphasising the filtering effects of primary and preventative health-care, whereas Figure 6.4 is a top-down model, indicating the critical importance of the tertiary level to the system. If a system like that of Figure 6.5 were to be functioning effectively (implementation has commenced), then the costs of the upper levels of Figure 6.4 would be in proper proportion and a better resource allocation would be achieved.

It is interesting in looking at Figure 6.5 to observe that engineers and architects fulfil a vital role in the lowest level of the model, i.e. supplying subsistence needs. The importance of these factors in promoting health and productivity is well known, and is closely connected with the wealth-creating effects of infrastructure mentioned earlier. Thus, the life cycle argument for

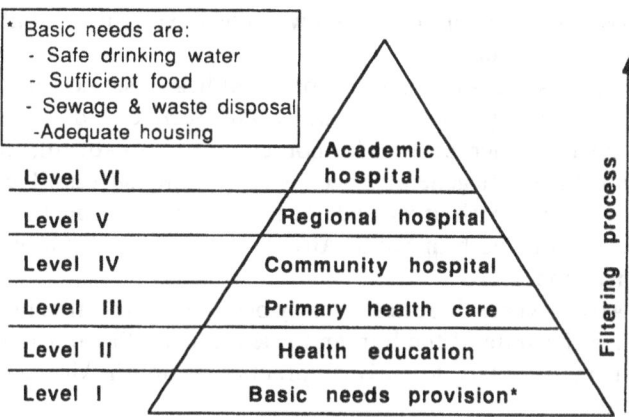

Figure 6.5 Multi-level health care system.

adequate infrastructural provision and maintenance is twofold—it promotes health as well as wealth.

6.9.2 Funding and payment

The second aspect of the system influencing resource utilisation is of less direct interest to construction professionals, but has parallels with the system of payment for design, which was mentioned earlier as a problem. Due to its labour-intensive nature, hospital care has shared in the general economic movement whereby manufactured goods have tended to fall in price in real terms, while the cost of services has tended to rise.

There is more to the issue than this, as hospital costs worldwide have risen faster than even the services cost trend, and any discussion of the reasons for this usually comes up against the question of the level of servicing provided. It is said by some authors that the system of paying fees for service to the medical profession encourages over-servicing (e.g. Broomberg and Price, 1990). However, this is only possible within the limits of the patient's ability to pay. Other researchers point the finger at the demand side, supported by the ability to pay created by insurance or medical aid. The patient naturally wants the best possible care, and so long as the patient does not pay directly, will not mind if extra or unnecessary tests or treatments are given (Feldstein, 1973).

Feldstein (1973) argued that 'the sensitivity of health-care demand to the extent of insurance coverage has been a major cause of the inflation of health costs'. Although some form of health insurance is desirable—'the inherent uncertainty of family medical-care costs creates a demand for health insurance'—Feldstein (1973) believes that insurance coverage is a mixed blessing. 'Because hospital care is more completely insured than other health services, insurance distorts the pattern of health-care towards the use of expensive hospital in-patient care even when less expensive ambulatory care would be equally efficient'.

The alternative system of a unitary national health service limits demand differently—by queueing. It is not unknown for recipients of national health services to wait a considerable time for non-emergency treatment, and national health systems have not generally been a resounding success. Despite these difficulties, disillusion with privatised, fee-for-service medicine runs deep, and there have been calls both in South Africa and the USA for more emphasis on public medicine.

In South Africa, where there has recently been a push towards privatisation, in order to rid the state of the burden of health care, Benatar (1990) argues that while there is a place for private practice in South Africa, this should not dominate:

'...further increase in the private-practice component will not help to cater for the health needs of the majority of the population, for education of health-care professions or for the survival of academic medicine. Privatisation and fee-for-service medicine should not be over-encouraged and, in particular, not to the detriment of public medicine.'

(Benatar, 1990)

A similar call comes from the USA.

'Our health-care system is failing. Tens of millions of people are uninsured, the costs are sky-rocketing and the bureaucracy is expanding. Patchwork reforms succeed only in exchanging old problems for new ones. It is time for a basic change in American medicine. We propose a National Health Program... The pressures of cost-control, competition and profit threaten the traditional tenets of medical practice. For patients, the misfortune of illness is often amplified by the fear of financial ruin. For physicians, it often gives way to anger and alienation... The world's richest health-care system is unable to ensure such basic services as prenatal care and immunisation.'

(Himmelstein and Woodlander, 1989)

If that is true for the 'world's richest health-care system', it is unlikely that South Africa with its mix of First- and Third-World components could do better. The debate illustrates the critical role of policy (top of the pyramid) not only in the success or failure of health-care provision, but also in whole life costing.

6.10 Conclusions

The foregoing discussion has focussed on the question of the allocation and effective utilisation of resources within the health-care system, and the effect of this on national productivity. The debate is, of course, part of a wider debate, concerning the allocation of resources within the economy as a whole.

The question of resource allocation is at root a life cycle problem, and a design problem of the highest importance, being at the very top of the pyramid model where policies are made.

'I believe that it is time to plan again in order to resolve the issues, that is the starting point. An efficient system is the cornerstone of a wealthy and civilised society. If we do not spend our way out of the problem, a poor and uncivilised society will become a self-fulfilling prophecy.'

(Kilvington, 1990)

The above quote refers not to the health-care system, but a country's transport system, and this is the heart of the dilemma. There are many worthwhile and important areas clamouring for resources (how well would a health-care system function without an efficient transport system?) and quality decisions are vital. A proper application of life cycle principles can make an important contribution to wise allocation of resources. Note, it is not the understanding of life cycle principles that is lacking, but the application, and this is due in large part to the systems which are in place.

Engineers and architects will have to become involved at the policy-decision level if there are to be worthwhile projects to design and resources to build them with. The same applies to medical personnel and the health-care system—nothing less than the quality of life of entire countries is at stake and that really is 'whole life costing'.

References

Benatar, S.R. (1990) A unitary health service for South Africa, *South African Medical Journal*, **77**, 444.

Bettigole, N.H. (1990) Replacing bridge decks, *Civil Engineering ASCE*, September, 76.

Broomberg, J. and Price, M.R. (1990) The impact of the fee-for-service reimbursement system on the utilisation of health services, *South African Medical Journal*, **78**, 130–132.

Business Round Table (1982) Integrating construction resources and technology into engineering, Business Round Table, New York, p. 8.

Civil Engineering ASCE (1990) A bioclimatic building, *Civil Engineering ASCE*, 71.

Construction Industry Institute (1986) Evaluation of design effectiveness, Construction Industry Institute, Austin, Texas.

Department of Health and Welfare (South Africa) (1981) National plan for health service facilities, Department of Health and Welfare, Pretoria, South Africa.

Engineering Council and Design Council (1986) Managing design for competitive advantage, The Engineering Council, London, p. 14.

Feldstein, M. (1973) The medical economy, *Scientific American*, September, 151–159.

Financial Times (1985) Dunlop takeover by BTR for £101 million—up from original bid of £33 million, *Financial Times*, 1.

Hahn, D.M. (1990) Consider life cycle costs, *Civil Engineering ASCE*, October, 30.

Heinen, R. (1985) Who should design bridges? *Civil Engineering ASCE*, July, 63–66.

Himmelstein, D.U. and Woodlander, S. (1969) A national health program for the United States, *New England Journal of Medicine*, **320**, 102–108.

Institution of Civil Engineers (1985) Congress discussion: Does professional competition benefit society? *Proceedings of the Institution of Civil Engineers, Part 1*, **78**(1), 589.

Kilvington, R. (1990) Address at Transport to Prosperity Conference, London, *Highways and Transportation*, June, 30.

Kirschenman, M.D. (1986) Total project delivery systems. *Journal of Management in Engineering*, **2**(4), 222–230.

Krizan, W.G. (1986) Institute pushing hard in productivity plan, *Engineering News Record*, 48.

Kuesel, T.R. (1990) Whatever happened to long-term bridge design? *Civil Engineering ASCE*, 57.

McGeorge, J.F. (1988) Design productivity: A quality problem. *Journal of Management in Engineering ASCE*, **4**(4), 350–362.

National Building Research Institute (1985a) Buildings for health and welfare services. *Research Information Newsletter*, National Building Research Institute, Pretoria, South Africa, Vol. 1, No. 4.

National Building Research Institute (1985b) Community health centres, *Research Information Newsletter*, National Building Research Institute, Pretoria, South Africa, Vol. 1, No. 3.

New Engineering Contract, The (1991) Need for and features of the NEC: in *The New Engineering Contract*, Thomas Telford, London, pp. 2–3.

Nicholson, J. (1991) Rethinking the competitive bid, *Civil Engineering ASCE*, January, 66.

Parkinson, C.N. (1960) *The Law and the Profits*, John Murray, London, p. 9.

Parkinson, C.N. (1967) *Left Luggage*, Penguin Books, Harmondsworth, p. 182.

Paulson, B.C. (1976) "Designing to reduce construction costs". *Journal of Construction Division ASCE*, **102**(104), 587–592.

Shepherd, D. (1987) Fees: A long hard look, *Engineering News* (South Africa), 22 May, 58.

Taylor, R.S. (1985) The influence of research and development on design and construction. *Proceedings of the Institution of Civil Engineers, Part 1*, **78**, 469–497.

Vlatas, D.A. (1986) Owner and contractor review to reduce claims. *Journal of Construction Engineering and Management ASCE*, **112**(1), 104–111.

Webb Committee of Enquiry (1975) Report of the Committee of Enquiry into norms and procedures for hospital construction, Department of Health and Welfare, Pretoria, South Africa, pp. 22–25.

World & I, The (1991) America's crumbling infrastructure: Losing wealth through the cracks (editorial), *The World & I*, **6**(8), 18–19.

7 How life cycle costing could have improved existing costing

A. ASHWORTH

7.1 Introduction

It has long been recognised that to evaluate the costs of buildings and engineering structures on the basis of their initial costs alone is unsatisfactory. Some consideration must also be given to the costs-in-use that will accrue throughout the life of the building or engineering structure. The use of life cycle costing for this purpose is an obvious idea, in that all costs arising from an investment are relevant to that decision. The image of a life cycle is one of progression through a number of phases, with the pursuit of an analysis of the economic life cycle cost as the central theme of the whole evaluation. While the proper consideration of the whole life costs is likely to result in a project that offers the client better value for money, there are problems which still need to be resolved before the method can be properly used in practice.

7.2 Review

The evidence for the use of life cycle costing in practice varies in different countries, and varies in the importance that is attached to long-term policies and priorities regarding the infrastructure and its buildings. The technique seems to have found more application in North America, according to Dell'Isola and Kirk (1981), Ahuja and Walsh (1983) and Jelen and Black (1983). This may stem from the different approaches and philosophies used for calculating the costs of construction work, which is very different to that used in the UK and in mainland Europe. Law (1984), however, states that the case is still being made in the USA for life cycle costing. Szoke (1986) suggests that, in Europe generally, the application of the technique is at about the same state as in the UK.

The technique, which is based upon discounted cash flow analysis, is not new, but has been borrowed from economic theory. It has been used in the investment appraisals of commercial and industrial activity for some time. The interest in life cycle costing in the construction industry dates back to the 1950s and to research undertaken at the Building Research Establishment on 'costs-in-use' (Stone, 1960). Those involved in the design of projects also became aware that their initial design solutions did have an impact upon the long-term economics of building usage. If the method of calculating total costs over the life of a project was widely understood and applied, the decisions about its design, the use of components and the choice of materials could have regard not only to their initial costs, but also to the consequences of such capital expenditure in terms of life expectancy, future replacement, repairs and other running costs. In 1974 the British Standards Institution published BS 3811, which describes the sequence of life cycle phases from specification through to eventual replacement, and while this adopts engineering terminology, the definitions used easily fit into the life phases of a construction project.

The increased interest in the use and development of the technique is indicated by the large amount of literature that has been published over the last twenty years. The Royal Institution of Chartered Surveyors (RICS) has consistently supported its projected use into practice. It commissioned a major study of the principles and its possible applications (Flanagan et al., 1983), and has since initiated further study through surveyors from the different specialist subgroupings within the RICS. This culminated in the publication of further reports in 1986 and 1987 (Royal Institution of Chartered Surveyors, 1986, 1987). The Society of Chief Quantity Surveyors in Local Government has prepared a report in the form of a practice manual (Smith et al., 1984). Architects on both sides of the Atlantic have also shown an interest in the technique, which resulted in publications by the American Institute of Architects (AIA, 1977) and the Royal Institute of British Architects (RIBA, 1986), as there are repercussions in the way in which they carry out their work. Hoar (1986), in endorsing the findings of the above reports, has suggested to practitioners that they will still need to convince their clients that life cycle costing is a worthwhile service that should be carried out. A confidential report from a North American source suggests that, in order to interest clients in applying the technique in practice, future costs may need to be exaggerated in order to make them look important. This tends to suggest that future costs can be made to 'say what we want them to say.' While this may not be the whole truth, as with every form of cynicism, there is an element of truth in this statement. It does nevertheless suggest that life cycle costs are capable of a wide range of interpretations. More recently, Flanagan and Norman (1989) have provided an update on the state of the art of life cycle costing and developed some of the more advanced applications.

While the principles of life cycle costing and the associated evaluative methods can be easily demonstrated in theory, there are difficulties in using the techniques in practice. These relate to a lack of knowledge and understanding, on the part of both practitioners and clients, and to a number of uncertainties, particularly in respect of historic data, the long-term future time horizons and the policy issues of asset management. There is also a feeling of vogue about the use of the technique, and that tomorrow it might be relinquished in favour of a more novel form of analysis. The attention, however, that it currently receives is a positive step forward for the construction industry.

7.3 The importance of forecasting

The importance of counting the cost before building was recognised at least 2000 years ago (St Luke's Gospel, 14; 28). The inference in this example is on the life cycle cost! Forecasting is required for a variety of purposes, such as early price estimating, the setting of budgets, invitation of tenders, cash flow analysis, final account predictions and life cycle costing. While it is recognised that there are confidence and reliability problems associated with initial cost estimating, these are not of the same magnitude as those associated with life cycle costing. A large amount of research has been undertaken in an attempt to improve the forecasting reliability of these methods. By comparison, the acquisition of life cycle costing knowledge and skills through research and application is still in its infancy, with a considerable gap between theory and practice. It is also difficult to provide confidence criteria, due largely to an absence of historical perspectives, professional judgement and a feeling for a correct solution.

7.4 Predicting the future

The fundamental problem associated with the application of life cycle costing in practice is the requirement to be able to forecast a long way ahead in time. While this need not be done in absolute terms, it must be done with sufficient reliability to allow the selection of project options that offer the lowest whole life economic solutions.

The main difficulties are similar to those of weather forecasting and other activities that attempt to predict a future event. All forecasts are fraught with some sort of confidence credibility, but this does not mean that they should not be attempted. Weather forecasting, for example, has access to huge amounts of data that have been systematically collected over time, and has also benefited from extensive research programmes. Even so, it is still unable to provide reliable forecasts of weather patterns for even a few days ahead. Life cycle costing by comparison is a social-science skill, where human perceptions and experiences including their vagaries are manifest, and where aspirations, objectives and desires fluctuate and evolve. It is in this context that life cycle costing is to be applied, and predictions, if they are to be useful, need to be reliable for at least the next quarter of a century.

The major difficulties that face the application of life cycle costing in practice are therefore related to predicting the behaviour of future events. While some of these events can, at least, be considered, analysed and evaluated, there are other aspects that cannot even be imagined today. These remain, therefore, outside of the scope of prediction and probability, and are unable to be even considered, let alone assessed in the analysis.

7.5 The common issues

During a life cycle cost analysis there are many essential facets that have to be determined, often on the basis of only scant evidence, data and information. Some of this information is of such a crucial nature that high-quality professional judgement, forecasting and insight are necessary if acceptable results are to be achieved. Factors such as the building's or structure's life, the lives of the various components used, together with their repair intervals and costs, the discount rates to be selected over the whole life cycle, the rates of interest, the amounts of inflation, and the influences of future fiscal policies all need to be predicted. A wide range of values can be allocated to all of these component parts of the analysis, which will result in diverse solutions being achieved. Ashworth (1987a) has outlined the fundamental principles associated with the particular issues relating to these factors. There are, in addition, other techniques that can be used to test the reliability of a life cycle cost analysis. Sensitivity analysis, for example, will allow for further evaluation of the results and provides for an increased measure of confidence in the predictions. Such techniques help to reduce the level of uncertainty, although this can never entirely be removed where the future is concerned. This enhanced analysis is still, however, an over-simplification of the behaviour of the real world of construction costs, and the model's reliability is capable of being distorted in many different ways.

Furthermore, we live in an age when many of these matters are of immense importance to owners, designers and users of buildings and engineering structures; but will this always be the case? Life cycle costing may become of even greater importance, but because of the realisation that the future is really unpredictable, it may be discarded in favour of other aspects of construction economy that can be more readily determined. Flanagan (1984) suggested that some of the issues have already been resolved, but some need further research before the technique can receive extensive practical application. This is still the case today.

7.6 Advantages of life cycle costing

It is worth noting, briefly, the existing methods that life cycle costing enhances or replaces. These range from the single-price methods of estimating which are used for predicting initial capital-cost budgets, to cost analyses and cost planning (Ashworth, 1987b). While the evolution and refinement of these methods has been extensive during the past twenty years, they all suffer from the inherent defect of evaluating projects on the basis of their initial construction costs alone. There is an absence of any consideration of the future costs associated with the use of the project. The emphasis of the past has therefore been towards improving the accuracy and reliability of early price forecasting (Ashworth and Skitmore, 1982) and the cost planning of the design process (Royal Institution of Chartered Surveyors, 1980) rather than towards the evaluation of whole life costs. An improvement on these methods, but nevertheless a rather simplistic and blunt method of accounting for both initial and future costs, is the payback method. This technique is widely used in practice, albeit for projects with short life spans, to determine financial cut-off points of competing proposals. This technique falls short in its application, since no account is taken of the timing of the cash flows or of the time value of money. The central theme of life cycle costing, however, is the attempt to redress these imbalances, by evaluating projects on the basis of a combination of initial and future costs. A life cycle cost approach, that is, an approach that takes explicit account of the life cycle costs of assets, is essential to effective decision-making in the following ways (Flanagan *et al.*, 1983):

(i) Life cycle costing is a whole or total cost approach undertaken in the acquisition of any capital-cost project or asset, rather than merely concentrating on the initial capital costs alone.

(ii) Life cycle costing allows for an effective choice to be made between competing proposals of a stated objective. The method will take into account the capital, repairs, running and replacement costs, and express these in consistent and comparable terms. It can allow for different solutions of the different variables involved and set up hypotheses to test the confidence of the results achieved.

(iii) Life cycle costing is an asset management tool that will allow the operating costs of premises to evaluated at frequent intervals.

(iv) Life cycle costing will enable those areas of buildings to be identified as a result of changes in working practices, such as hours of operation, introduction of new plant or machinery, use of maintenance analysis, etc.

7.7 Life cycle costing applications

The following are some of the main applications of life cycle costing associated with construction projects.

7.7.1 At inception

Life cycle costing can be used as a component part of an investment appraisal. This is the systematic approach to capital investment decisions regarding proposed projects. The technique is used to balance the associated costs of construction and maintenance with rental values and needs expectancies. It is a necessary part of property portfolio management. It recognises that many projects are built for investment purposes. The way that future costs-in-use are dealt with therefore largely depends on the expected ownership criteria of occupation, lease or sale, or indeed a combination of these alternatives.

7.7.2   At the design stage

A main use of life cycle costing is at the design stage or precontract phase of a project. Life cycle costing can be used to evaluate the various options in the design in order to assess their economic impact throughout the project's life. It is unrealistic to attempt to assess all the items concerned; indeed, the cost of undertaking such an exercise might well rule out any possible overall cost savings. The sensible approach is to target those areas where financial benefits can be more easily achieved. As familiarity with the technique increases, it becomes easier to carry out the analysis, and this may prompt a more in-depth study of other components or elements of construction. While some of the areas of importance will occur on every project, others will depend on the type of project being planned. For example, roofing is probably an important area for life cycle costing on most projects, whereas drainage work is not. However, on a major highway scheme, where repeatability in the design of the drainage work occurs, then the small savings that might be achieved through life cycle costing can be magnified to such an extent to make the analysis worthwhile. The important criterion to adopt is that of cost sensitivity in respect of the whole project costs.

Life cycle costing is perhaps most effective at this stage in terms of the overall cost consequences of construction. It can be particularly effective at the conceptual and preliminary design stage, where changes are able to be made more easily, and where the resistance to such changes is less likely, than when a design is nearing completion. In these circumstances the designer may be reluctant to redesign part of the project even though long-term cost savings can be realised.

In selecting a design from a possible choice of options, the choice with the lowest life cycle cost will usually be the first choice, provided that other performance measures or criteria have been met. Using life cycle costing with other techniques, such as value engineering, should enable the scheme to be designed within a framework that is more cost-effective without the loss of any of the design's desirable attributes.

7.7.3   At the construction stage

While the major input of life cycle costing is at the design stage, since its correct application here is likely to achieve the best in overall long-term economic savings, it should not be assumed that this is where the use of the technique ceases. At the construction phase there are three broad applications which should be considered.

The first of these concerns the contractor's method of construction, which, unless prescribed by the designer, is left to the contractor to determine. In some instances the contractor may be allowed to choose materials or components that comply with the specification but will nevertheless have an impact upon the life cycle costs of the project. The method of construction which the contractor chooses to employ can have a major influence upon the timing of cash flows and hence the time value of such payments. This is perhaps more pertinent to works of major civil engineering construction, where the methods available are more diverse. Buildability aspects that might enable the project to be constructed more efficiently, and hence more economically, may also have a knock-on effect in the longer term and hence have an influence upon the related costs-in-use.

Secondly, the contractor is able to benefit from adopting a life cycle costing approach to the purchase, lease or hire of the construction plant and equipment. The probable savings resulting from this evaluation may then have an impact upon future tendering and estimating strategy and project costs.

Thirdly, the construction managers are able to provide a professional input to the scrutiny of the design, if involved sufficiently early in the project's life. They may be able to identify life cycle cost implications of the design in the context of manufacture and construction and in the way that the project will be assembled on site.

7.7.4 During the project's use and occupation

Life cycle costing has an important part to play in physical asset maintenance management. The costs attributable to maintenance do not remain uniform or static throughout a project's life. Maintenance costs therefore need to be reviewed at frequent intervals to assess their implications within the management of costs-in-use. Taxation rates and allowances will change and these can have an impact upon the maintenance policies being used. Grants may also become available for building repairs or to address specific issues such as energy usage or environmental considerations. The changes in the way the project is used and the hours of occupancy, for example, all need to be monitored to maintain an economic life cycle cost, as the project evolves to meet new demands placed upon it.

When a project nears the end of its useful economic life, careful judgement needs to be exercised before further expenditure is apportioned. The criterion for replacing a component is a combination of the rising running costs compared with the costs of its replacement and its associated running costs. Additional non-economic benefits are also considered and need to be accounted for in the analysis. For example, the advancement made in the improved efficiency of central heating boilers and their systems suggests that these, on economic terms alone, should be replaced every 10–12 years irrespective of their working condition. A simple life cycle cost analysis is able to show that the improved efficiency of the burners and the better environmental controls will outweigh the replacement costs within this period of time.

7.7.5 At procurement

The concept of the lowest tender bid price should be modified in the context of life cycle costing. Under the present contractual and procurement arrangements, both manufacturers and suppliers are encouraged to supply goods, materials and components which ensure their lowest initial cost irrespective of their future costs-in-use. In order to operate a life cycle cost programme in the procurement of capital works projects, greater emphasis should be placed upon the economic performance in the longer term, in order to reduce future maintenance and associated costs. The inconvenience that often arises during maintenance and the other associated replacement costs, which may be out of all proportion to the costs of the part that has failed, also need to be examined. The different methods of procurement that are available may also make it easier and more beneficial for the contractor to consider the effects of life cycle costing on a design.

According to Ahuja and Walsh (1983), the US Federal Supply Services, which have operated a form of life cycle cost procurement for a number of years, feel that they are achieving considerable success in applying this technique and reducing whole life costs.

7.7.6 In energy conservation

A major goal of the developed nations is towards a reduction of energy use in all of its costly and harmful forms. This is true for the governments concerned, who have introduced taxation penalties, and for private industry, who are seeking ways of reducing their own energy consumption and hence the associated costs. Life cycle costing is an appropriate technique to be used in the energy audit of premises. A reduction in energy usage has been encouraged, due to the rising costs of foreign oil supplies, the finite availability of such fossil fuels, and what has now become commonly known as the 'greenhouse effect'. The energy audit requires a detailed study and investigation of the premises, recording of outputs and other data, tariff documentation and an appropriate monitoring system. The way the premises are used, plus typical or likely expectations of energy usage, and sound professional judgements are important criteria for such an analysis. The recommendations may include, for example, providing additional insulation in walls and roofs, and the replacement of obsolete equipment, as well as suggesting values for temperature gauges, thermostats and other control equipment. An energy audit is not simply a one-off calculation, but one that needs to be repeated at frequent intervals in order to monitor the changes in the variables that affect the overall financial implications.

7.8 Kinds of uncertainty

There are a number of major difficulties that influence the utilisation of life cycle costing in the construction industry. Forecasting can only be carried out in the light of present knowledge. The future can only be predicted within the limits of present-day expectations.

7.8.1 Life expectancy

Traditionally, most commentators on life cycle costing have based their calculations on a sixty-year life span of buildings. This originated from a development surveyors's perspective, associated with rentals and yields, and from the technologist's view regarding construction longevity and obsolescence. Can we realistically and honestly hope to forecast costs for sixty years ahead? It is now generally accepted that the life cycle time horizon should be increasingly related to current use expectations associated with the building's structure's own cycle, or related to the cyclical effect of population movements associated with the project. This might be reflected, for example, in a twenty-year life cycle for school buildings. Even then calculations, in a rapidly changing world, may be wide of the mark.

7.8.2 Data difficulties

A frequently held reason for why the technique has not been more widely used is the lack of appropriate, relevant and reliable historical cost information and data. Maintenance cost data have, in the past, largely been collected solely for accounting purposes and to satisfy the reconcilation of budgets. These accounting headings are often unsuitable for use in life cycle costing applications. Where data have been found to be available they are often so contradictory in their nature, that their satisfactory reuse becomes almost impossible. Holmes and Droop (1982) illustrate this, after analysing a large amount of maintenance data from local authority projects such as housing and schools. While the mean values of the data offered some consistency their standard deviations exhibited considerable fluctuations. An understanding of the contextual nature of the data is very important where the reuse of any of them is envisaged. Fletcher (1990) was able to obtain and analyse a large amount of data from local authority housing schemes. This study further emphasised the mismatch between the cost headings used by maintenance departments and those required for use in the life cycle costing of future projects.

The data used for estimating by contractors for tendering purposes have wide variations in outputs but these are relatively insignificant when compared with the inconsistency of maintenance cost records. Furthermore, little information is provided of a qualitative nature; this is required if the costs-in-use data are to be properly interpreted. The broad categories of accounting headings such as 'general maintenance' or 'repairs' disclose too little information to prospective data users on which sound judgements of reliability can be made.

Those working in practice are also waiting for others, particularly those with access to possible huge data sources, to provide databanks, before proceeding further with putting the principles into practice. Although there are capital-cost databanks for initial price forecasting, such as estimating outputs, 'standard' prices, cost analyses and other supporting information, these are only used as a source of second opinion, or where a practitioner's own data either do not exist or are deficient in some other way. The application of historic cost records to new projects requires a good deal of judgement if the desired results are to be achieved. Maintenance-cost data are of a more critical nature than this, due to the other influences which affect their quality and reliability. The inherent characteristics of such data must be known to the user, and any published databank of maintenance costs may therefore be of limited real use to the practitioner. For example, it is necessary to have some insight into the causes of component failure or deterioration. Was the repair carried out due to a failure of other items of work, vandalism, misuse or simply normal wear and tear? Furthermore, the lives of components or materials may have become shortened due to the time-lag occurring between reporting, remedy, invoicing and payment stages. The historic information can therefore easily be misleading in terms of both component life and its attributed costs. In capital-cost estimating, the practitioner's own database is of paramount importance; with maintenance-cost data this is even more true.

7.8.3   Technological change

It is difficult to forecast with any degree of accuracy the possible changes in technology, materials and construction methods that may occur over the next decade. The construction industry, its process and its product are under a purposeful change and evolution. There is a constant striving to develop excellence in both design and manufacture and to introduce new materials having the desired characteristics of quality and reliability in use. The changes in technology can often be sudden and unexpected. Prototypes that when used in practice may fail initially are eventually refined and improved to produce a worthwhile product. The introduction of new technology and good solutions to age-old problems can have a major impact on the life cycle cost forecasts and in the pursuit of whole life construction economy.

7.8.4   Fashion changes

A further difficulty facing the application of life cycle costing in practice is changes in fashion. These changes are less gradual and more unpredictable than changes in technology and are also subject to a degree of speculation. Themes within the construction industry have been developed in different eras; examples are 'built to last', 'inexpensive initial cost', 'industrialisation', and 'long life, loose fit, low energy' (Gordon, 1974), and the present attitudes towards refurbishment. Changes, for example, in the type and standards of provision, the use of space or the level of quality expectations can be observed from the past study of buildings. Changes in the way that buildings might be used in the future are already predicted. Some of these are hopelessly fanciful. Others reflect an attitude to work and leisure, changes in the individual's personal expectations, demographic trends and developments generally in society. A life cycle cost analysis must, however, attempt to anticipate future trends and their future effect on the overall economic solution. Fashion changes are the result of the desire to provide something new, sometimes solely to address a reason for change. In other cases they arise due to our social awareness and perception of human development and advancement. A life cycle cost analysis which considers only the status quo is of very limited value in practice.

7.8.5   Cost and value changes

The erratic pattern of inflation throughout the past twenty-five years could not have been predicted even a decade earlier. The high inflation experienced during the 1970s would not have been thought possible in the 1950s. An examination of building tender prices throughout the 1960s and 1970s indicates a general upward trend in the values of these data. This pattern has existed since the end of the depression of the 1930s. In the early 1980s, however, tender price levels showed a downturn, which at the time was an unusual and unexpected phenomenon, since the preceding years had already been financially difficult times for builders and contractors. The more recent variability of oil price levels illustrates how volatile the market-place really is. Inflation rates and interest rates are intertwined and influenced by such factors. Slumps follow booms and *vice versa*, but even so, these seem to be beyond the scope of present indicators and predictors. Costs and values do not move in tandem; neither do the respective indices for the different materials, products or components follow similar patterns but are subject to wide degrees of fluctuations. Economists have indicated that costs and prices cannot be expected to rise indefinitely, and that there may be a future lapse or even a reversal of the traditional historic patterns.

7.8.6   Policy and decision-making changes

One of the most important life cycle costing variables is the future use and maintenance policy of the project by the owner. This factor is also the one that is generally absent from the sparse historic data sources that are available. It is now widely recognised, for example, that maintenance work is not needs-oriented but budget-led. The maintenance work that is carried out is thus largely determined by the amount of funds available. Once these funds have been expended then no further amounts are available until the following year's budget allocations have been determined. There may therefore be only limited value in comparing the whole life costs of, say, wall tiling with those of repainting, in the absence of such a policy. The tiling may be shown to be the economic choice, but if the owner, due to a shortage of available funds, does not repaint the walls at the intervals that have been stipulated in the life cycle cost plan, then the economic comparison may prove to have been at best optimistic or even a false assumption. The policy of the owner and the use by the occupants are likely to be characteristics at least as important as the theoretical design and construct values in the determination of the relevant maintenance costs.

The way in which owners and occupiers use and care for their buildings or other structures also needs to be considered. The desire for proper maintenance of the physical asset is influenced by the costs and inconvenience involved. Different owners will also set differing priorities. They cannot be assumed on the historical precedents of apportionments of other buildings unless it is certain that the uses and priorities are compatible. A study by the NBA (1985) suggested that maintenance cycles and their associated costs must firstly be set properly within the maintenance objectives of the particular organisation concerned and the policies employed for planned and responsive maintenance.

7.8.7   Accuracy

One of the main criteria in any estimate is its reliability or accuracy. By definition an estimate will never be spot-on. The inaccuracy of capital-cost estimating in the forecasting of contractor's tender sums has been measured to be about 13% (Ashworth and Skitmore, 1986). Contractors' estimating of their own costs is marginally better. The processes used for both of these types of forecast have been refined through many years of use, experience and 'feel' in practice. Life cycle costing is relatively new with limited experience in practice and a quality of data that is very subjective and inferior to that used for capital-cost estimating. The reliability of the results achieved will be subject, therefore, to much larger variations and possible errors than those indicated above for capital-cost estimating. The key criterion, however, for life cycle costing is the accuracy of the comparability of design options in allowing the correct economic solution to be made. This, however, must be made in the knowledge of large possible estimating inaccuracies.

7.9 Historical perspectives

The presumption is often made that life cycle costing will assist in the selection of the most economic solution for a design, taking into account all of the costs associated with that project. A brief precursory glance at the past suggests that this may not always be the case. Consider, for example, a simple exercise concerned with the evaluation of timber and cast-iron rainwater gutters, which might have been made at the beginning of the twentieth century. Cast iron would have been selected as the economic solution, largely because of its durability and low costs-in-use when compared with timber. However, within a few years of such a decision being taken, a new material now known as PVC had been discovered for use in gutters. The correct economic solution based upon hindsight and historical fact would have been to have installed the timber gutters and when replacement became necessary to have renewed them with PVC.

Flat roofs are out of fashion today, primarily because of their apparent short life, high repair cost when compared with pitched roofs, and their low reliability. Life cycle cost calculations do not generally favour them, even under the most optimistic conditions when compared with an inexpensive pitched roof construction. The recommendation today, therefore, after all of the economic considerations have been examined, is to choose the latter. However, it is possible that, within a few years, materials scientists may discover or invent a material for flat roofing that is inexpensive, highly durable and reliable and has a life expectancy and costs-in-use that are lower than those of even moderately priced pitched roofs. The correct economic choice may therefore be to install the cheaper alternative flat roof construction, and then replace this after its normally expected short life with this, yet to be discovered, material.

The provision of insulation in buildings is a reflection of the relationship between the annual cost of fuel for heating purposes and the initial cost of the insulation. The search in recent years for alternative and less expensive forms of fuel has been an ambition yet to be realised. When these are discovered, much of the present levels of insulation in buildings may become redundant in terms of their cost effectiveness. The real reduction in the price of fossil fuels and other energy sources in recent years, together with the added efficiency of mechanical heating plant and equipment, provides this argument with some validity.

The illustrations above, which range from fact to fiction, indicate that it is possible to use life cycle costing to help us demonstrably to select the wrong economic option in a total cost appraisal. It is also worth noting that such a choice could even have been made in ignorance of this technique! The question, however, that needs to be answered is, if the technique is applied to projects constructed tomorrow, will the long-term desired objectives be achieved? The technique does not remove from users the responsibility to apply judgements and to make decisions, but it needs to offer a reliable analysis on which to base these decisions.

7.10   Conclusions

The importance of attempting to account for future costs-in-use in an economic appraisal of any construction project has already been established in theory. The question of whether life cycle costing works in practice is of crucial importance, and this is perhaps best assessed from models using evidence of past performance. Ashworth (1987a) has set out a procedure to test the generally held hypothesis that life cycle costing can become a practice tool in the construction industry. The general belief is that life cycle costing when applied to capital-works projects will enable the selection of the most economic solution over the project's whole life. This might not be so. If it can be shown, for example, that life cycle costing might have encouraged the choice of the least economic alternative, then its continued use in practice and its further development become of questionable worth for clients and their practitioners. If the forecasts are unreliable because of an absence of appropriate data then this should be a problem that can be remedied, at least in the long term, by properly assembling the datasets with the appropriate characteristics. If estimates are misleading because they rely upon the myth of being able to forecast the future, then the efforts in evaluating alternative designs and methods of construction might be better spent in considering other more suitable techniques. The world is now undergoing very rapid change where new technologies are affecting all aspects of society. The present values in society are also under constant scrutiny and evolution, and it is virtually impossible to predict how these factors might influence the future. What is certain is that these aspects do have an effect upon life cycle cost predictions. In the past some of these would have been at the best misleading.

Life cycle costing does offer some potential. Its philosophy of whole-cost appraisal is certainly preferable to the somewhat narrow initial-cost estimating approach. The widespread effort so far expended in its research and development is a positive move; however, more research is necessary to sharpen up the realities of the problems encountered. There is also sometimes an eagerness to introduce a new method of evaluation without being fully aware of all of the facts. Improving the education of those who are responsible for the design of capital-works projects, and encouraging them to consider the future effects of their design and constructional details are urgent priorities. Educating owners and users in how to obtain the best out of their buildings is another useful course of action. The implementation of maintenance manuals or building owners' handbooks might also provide an improvement in the performance of buildings in use. At this stage, it is doubtful whether too much emphasis or importance should be placed upon the actual numerical results, due to the vagaries within the calculations. Although the use of this technique in practice will hopefully continue to increase for the reasons described above, this must be done with some caution, until results achieved in practice can be verified.

Life cycle costing is at best a snapshot in time, in the light of present day knowledge and practice, and anticipated future applications. Some of the factors involved are of a crucial nature and can only be tested over a range of known values. Others are currently beyond our expectations, may not even be considered as being important today, and may not come to light until observed in practice at some time in the future. Some of the assumptions may also be realised as untenable in practice. Are we asking too much of the technique?

References

Ahuja, H.N. and Walsh, M.A. (1983) *Successful Construction Cost Control*, Wiley Interscience.

AIA (1977) *Life Cycle Cost Analysis; A Guide for Architects*, American Institute of Architects.

Ashworth, A. (1987a) Life cycle costing—can it really work in practice? in: *Proc. Building Cost Modelling and Computers*, University of Salford, pp. 569–576.

Ashworth, A. (1987b) *Cost Studies of Buildings*, Longman Scientific and Technical, Harlow.

Ashworth, A. and Au-Yeung, P. (1987) The evaluation of life cycle costing as a practical tool during building design, in: *Proc. Fourth International Symposium on Building Economics*, Vol. A, Copenhagen, pp. 82–93.

Ashworth, A. and Skitmore, M. (1982) Accuracy in estimating, Occasional Paper No. 27, Chartered Institute of Building, London.

Ashworth, A. and Skitmore, M. (1986) Accuracy in cost estimating, in: *Proc. Ninth International Cost Engineering Congress*, Oslo, Paper A3.

Dell'Isola, P.E. and Kirk, S.J. (1981) *Life Cycle Costing for Design Professionals*. McGraw-Hill.

Flanagan, R. (1984) Life cycle costing—the issues involved, *Proc. Third International Symposium on Building Economics*, Ottawa, VI, Paper A1.

Flanagan, R. and Norman, G. (1989) *Life Cycle Costing: Theory and Practice*, BSP Professional Books, Oxford.

Flanagan, R., Norman, G. and Furbur, D. (1983) *Life Cycle Costing for Construction*, Surveyors Publication.

Fletcher, J. (1990) Housing maintenance costs analysed, MSc thesis. University of Salford.

Gordon, A. (1974) The economics of the 3L's concept, *Chartered Surveyor Building and Quantity Surveying Quarterly*, June, 31–36.

Hoar, D. (1986) Life cycle costing, *Chartered Quantity Surveyor*, 9(2), 37.

Holmes, R. and Droop, C. (1982) Factors effecting maintenance costs in local authority housing, in: *Proc. Building Cost Techniques*, Portsmouth Polytechnic, pp. 398–409.

Jelen, F.C. and Black, J.H. (1983) *Cost and Optimisation Engineering*, McGraw-Hill.

Law, A. (1984) Life cycle costing in the USA, *Chartered Quantity Surveyor*, 6(9), 338–339.

NBA (1985) Maintenance cycles and life expectancies of building components and materials, National Building Agency Construction Consultants.

Orshan, O. (1980) Life cycle cost: A tool for comparing building alternatives, in: *Proc. Symposium on Quality and Cost in Building*, Vol. 1, Laussane, pp. 52–71.

Royal Institute of British Architects (1986) Life cycle costs for architects, a draft design manual, College of Estate Management.

Royal Institution of Chartered Surveyors (1980) *Pre-contract Cost Control and Cost Planning*, Surveyors Publications.

Royal Institution of Chartered Surveyors (1986) *A Guide to Life Cycle Costing for Construction*, Surveyors Publications.

Royal Institution of Chartered Surveyors (1987) *Life Cycle Costing: A Worked Example*, Surveyors Publications.

Smith, G. *et al.* (1984) *Life Cycle Cost Planning*, Society of Chief Quantity Surveyors in Local Government.

Stone, P.A. (1960) *Economics of Building Design.*, Building Research Establishment.

Szoke, C. (1986) Present objectives and potential new vistas of working commission, W55 Building Economics, in *Proc. CIB*, Washington, 12–18.

8 Life cycle cost analysis: a decision aid
J.J. GRIFFIN

8.1 Introduction

When we embark on a project we should be aware that a large proportion of its total cost will occur during its in-service life, typically from 50% to as much as 80%. Thus, the earlier design, development, construction and manufacturing activities may be as little as 25% of what we will subsequently need to operate, maintain and overhaul our new asset (Figure 8.1). This simple statistic shows just how important it is at the initial concept, design and authorisation of a project, to consider the resources that will be needed to operate it throughout its life. These manpower and material resources all translate directly into money. The initial decisions on the project and its detailed design are going to lead to continued, and largely unavoidable, expenditure over a period of many years. An even more salutary observation is that few, if any, of the early decisions can be changed later except by the expenditure of even more money, possibly accompanied by temporary loss of facility.

All too often project authorisation has been based only on first cost with the acquisition authority or developer paying scant attention to their own or their successor's future cash flow. Fortunately there is a growing awareness of this situation and recognition that by making the correct decisions, an organisation can have a significant influence on its own future expenditure, or a developer can enhance the value of the investment to an operator or tenant. This chapter considers some of the techniques available for life cycle cost analysis (LCCA).

8.2 Project decisions

In project work we are continually faced with choosing from among a number of competing options. Each option has a number of attributes that must be taken into account in the decision. The principal attributes are performance, timescale or timeliness and cost, with 'performance' probably being expressed as some combination of a number of technical and operational characteristics which together define the requirements that we seek to satisfy. The detailed techniques for cost-benefit analysis are beyond the scope of this book, but

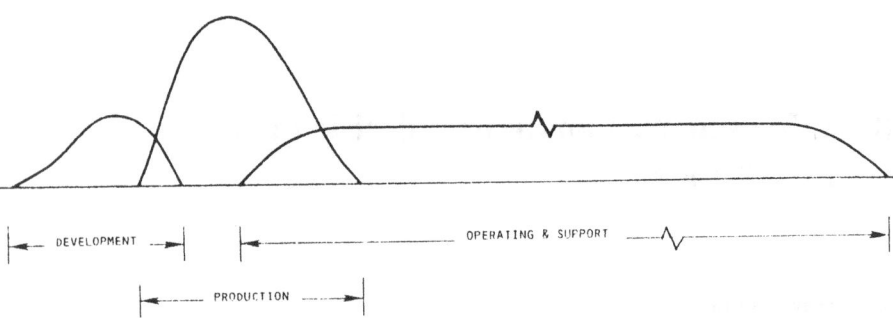

Figure 8.1 Life cycle costing profile.

the concept of analysing and quantifying options is central to this treatise. Life cycle costing is involved with the calculation of the likely cost of acquiring and then operating or owning an asset for the whole of its planned or useful life. Life cycle costing is a decision support tool. During the early phases of the acquisition process, during requirements, feasibility or option studies, the course of action which is most attractive or beneficial is quantified in an attempt to make the decision process as informed and numerate as possible. This is not to suggest that the decision itself is any easier; there is inevitably a compromise and trade-off between cost and performance since performance probably cannot be afforded at any price and it is rare for the best also to be the cheapest. There will also be cost–time and performance–time trade-offs.

However, cost does not exist as an independent entity. Cost arises only as a consequence of consuming some resource or asset which must be paid for or whose value is denied to some other use. Cost is a dependent variable that can only be measured or forecast in terms of the resource entity. Thus, the life cycle costing work is an attempt to 'model' the acquisition and operating processes in terms of the resources consumed and to convert all these resources to a single baseline cost total and cost profile. The conversion to cost is largely the relatively straight-forward arithmetic process of applying costing factors to all these material and manpower resources.

The ability to perform life cycle cost analyses that will be useful for management decision making is largely dependent on the ability to predict the amount and timing of future resource consumption. The time dimension is often important in cost calculations: the time when resources are bought and the rate at which they are consumed can have a significant cost influence. Time or the schedule of a project's work can, if it is not properly managed, have a dramatic influence on cost: schedule overruns can be very expensive even though the amount of useful work done changes very little.

If the 'management decision' is the personal one of selecting from among a number of similar consumer products, the life cycle analysis is a comparatively straightforward activity. The published prices, performance characteristics and facilities of each option will be well documented, and often there will be some 'reliability' information from consumer magazines, 'informed' experience or the perceptions created by advertising, in addition the performance attributes deemed to be essential or most highly valued will be known and a budget, a ceiling price or worth will have been set.

It is in decision making and option selection on larger and more expensive projects that a degree of innovation is possible (or uniquely necessary) and where a range of options exists that life cycle cost analysis is needed most. This innovation leads to high performance and cost risks. In such situations the relative clarity of the consumer product decision is no longer available. The project options may be a number of different ways of satisfying the requirement with different technologies and materials, at different locations or on a different scale. The options may also be major variations within an overall concept covering operating regime, commercial and procurement arrangements, support methods, maintenance and support policy or schedules.

The requirements that must be satisfied will include those of an explicit performance nature (e.g. the throughout of a process plant) as well as those of a more general nature such as national and local government regulations and industry standards. Such considerations as planning, new legislation and pollution may also affect the analysis. Even less tangible 'aesthetic' attributes such as architectural design and landscaping may be important requirements: they can certainly have significant cost consequences.

These requirements must be well enough established to allow the costing of viable and complete alternative solutions. There is no point in costing a non-compliant, non-feasible or unacceptable option.

Often an organisation will be considering the replacement of an existing facility. In such circumstances it is often useful to cost the do-nothing option of updating the facility currently in service. It will certainly be instructive. Running on the existing system or modifying it may seem inexpensive in first cost terms, but may generate high operating costs to support obsolescent technology. At the very least the do-nothing analysis gives a basic marker to the study and a valuable basis for comparison of all the other options.

Though the principal purpose of LCCA is as an aid to decision making and the selection of alternative designs, operating procedures or procurement strategies, it is also valuable in predicting future cash flow. The budgeting benefits that this brings are obvious and important but will be addressed no further here: instead management decision making and the selection of project options will be covered.

8.3 Timing

Life cycle cost analysis may be conducted at any time during the life of a project. The most beneficial time is during the early viability, feasibility or project definition phase of a project. It is at this time that most, if not all, options are open. The decision maker has maximum freedom and is unconstrained by existing commitments and decisions. It is the time when there is maximum influence over the future outcome and cost of the project.

8.4 Depth

The level of detail at which a life cycle cost analysis is carried out is set by:

(a) the level of the decision being made
(b) the level of information available.

It is clearly a waste to try to conduct the analysis at more detail than the decision needs, but neither should the work be inhibited by a lack of determination in obtaining data. In fact the study should be a mechanism for stimulating thought and the quantification of possible future project circumstances.

8.5 Refinement and evolution

The two strands of updating the LCC analysis as better information becomes available and as the level of decision becomes more detailed, lead to the concept of the evolution and refinement of the LCC 'model' as the project progresses.

The LCC undertaken in the early stages of the project can be updated as better information and more assured data become available during the subsequent development and acquisition stages. These enhancements at finer levels of detail and with less uncertainty can be undertaken at any subsequent decision point or milestone. The analyses will confirm the early decisions or lead to timely responses to new circumstances or proposed changes to the project. The evolving LCCA should form part of the on-going project management and review processes.

8.6 Scope

One of the most important and difficult decisions on any LCC study is to set the scope of the work. What elements of the system should be included? Clearly everything that affects the cost of the project must be taken into account.

This is a comparatively straightforward process if one element within an existing system is being changed or replaced, e.g. the control equipment for an office heating and conditioning plant, all other conditions such as operating cycles and performance standards remain equal and the building itself remains unaffected except for the installation of some new sensors. It would become more complex if a change of these standards (quality change) was being considered where trade-offs of cost and effectiveness are needed or a change in the fuel used (tariffs now and in the future, thermal conversion efficiency, etc.). It becomes more difficult if relatively novel heating systems using new energy technologies (solar systems, heat pumps or CHP) are considered, where the trade-off of capital investment and running costs has to be explored and each option generates different expenditure profiles and timescales.

Still more difficult are projects where there are fundamentally different ways of working, for instance in the manufacturing and chemical industries where production quantities and rates can determine the technology choice and electronic systems where device yields have a crucial commercial influence. In such areas, technical factors are clearly very important in the decision process and the LCC analysis must be responsive to the real technical issues and the technical viability of the project.

Some very large governmental and international projects become even more difficult to evaluate with their multitude of technology, resource, finance and infrastructure questions.

Thus, the scope of a study can vary dramatically depending on whether we are comparing apples with apples or apples with oranges. The greater the difference in system concept, alternatives or complexity (the more 'oranges') the wider the system boundaries and the more complex the study becomes.

There are two fundamental considerations when setting the scope:

(a) What is the question we are helping to answer, what decision is to be made?
(b) What is the narrowest system boundary at which the alternative systems have no different cost influence?

8.7 Life

The service life of a project is the period from an equipment entering service (having completed its acceptance after development, completing commissioning or the delivery of first production item) to its withdrawal from service (de-commissioning, start of phase-out of equipment or close of production). However, the complete life cycle additionally includes the preceding build or design and development, and commissioning activities and thus covers a longer period than service life.

The service life may be set by the inherent durability of the equipment, an obsolescence, a required service period, or by commercial criteria such as investment or financing duration or pay-back period. The forecasting of durability or obsolescence is difficult and our studies should reflect these uncertainties.

If the system has to meet a specified service period and its durability is expected to be inadequate, its re-furbishment or overhaul will need to be costed at some time during the service life.

Some organisations use a standard service life for all their LCCA as a consistent measure of cost. This takes no formal account of the benefits accruing from a system whose natural life exceeds this period. Any extra life is effectively regarded as an unquantified bonus that will accrue after the fixed term.

8.8 LCC content

An LCCA must, by definition, consider all aspects of project work from cradle to grave. Usually this is work that remains to be done at the time the study is being conducted, though it may sometimes be of interest to conduct a historical survey of past events, decisions and expenditure that have not been previously recorded.

The earliest that LCCA work is undertaken is usually at the requirements phase when a desire, need or possibility is converted into a set of project objectives. These objectives cover technical performance, timescales and financial targets.

Major LCC studies will be carried out during the feasibility phase, when viable solutions to the requirements are identified, assessed and compared with one another, and broad trade-offs made between performance, time and affordability.

The LCCA should continue in support of the specification phase and then during the subsequent design phase. This whole life, total system treatment of a project in the early phase of a project, when all its attributes and their inter-dependencies are considered, is very important as a foundation for all the future decisions and activities. Once the design is complete there will be a definitive baseline LCCA suitable for project budgeting.

Further LCCA should then be carried out to refine the budgets as better information becomes available during development, prototyping, pre-production and introduction into service phases on 'production' projects, or during construction and commission on 'build' projects.

Finally the operations and support part of the life cycle (covering maintenance, support, spares and replacement) should be refined and improved as in-service experience is gained for better operating budgets and cost control.

The LCCA, even in the very early stages of a project, must consider all the above phases and include production, test, evaluation and training and the procurement of project specific items such as test equipment, initial spares holdings and replacement parts.

8.9 Treatment of system costs

8.9.1 Inflation

It is usual in LCC studies to ignore inflation and to work in constant fiscal units, usually those prevailing at the time of the study. Allowances for inflation are usually made when budgets are being prepared and future currency outflow is being assessed.

8.9.2 Discounting

Discounting is used to recognise the time value of money: today's expenditure is more important than tomorrow's. This is, of course, no justification for ignoring life cycle costs and making project decisions on first cost only. Discounted cash flow calculations are needed when the expenditure profile of one option differs significantly from that of another.

8.9.3 Sunk costs

The term 'sunk cost' often features in LCC studies. The usual definition is 'a cost from which benefit is no longer gained'. This is a useful definition which applies to such things as previous generation projects, research, pilot projects and technology demonstrators. In fact, the experience gained from them is immensely valuable to a new project and can have a significant influence on its cost. It is just that these benefits are available without further direct expenditure. They can be better defined as 'no-cost project assets'.

8.9.4 Inherited investments

A more difficult analysis problem is created by assets which have been acquired in the past but which are necessary to the successful operation of the new project or system. Examples of such factors are infrastructure elements which may already be in place for one system option and not for another. Such things as transport, easy access to raw materials and services or a skilled workforce can be very important.

Clearly the option or site without the asset will have to invest in acquiring it. It would, however, be wrong to assume that an existing asset is available at no cost: there may be new charges levied, re-training or new commercial arrangements to be made. Also the remaining life of an inherited asset may be less than that of the project and some allowance may have to be made for its replacement or re-furbishment, or there may be new charges for its use.

In summary care must be taken in assuming that essential project assets are free. This illustrates the importance of setting the system cost boundaries correctly to achieve true comparisons.

8.9.5   Disposals

There may be costs incurred in scrapping or de-commissioning a system. These are often ignored because the costs are relatively trivial or are similar from option to option. In any case disposal costs are unlikely to influence the decision to initiate the project. Only where the de-commissioning costs are known to be significant and unavoidable (e.g. a nuclear power station) and/or related to some but not all system options should disposal costs be taken into account.

There may even be an income to be derived from selling the system. Income potential is usually ignored since it is difficult to put a market value on something a long way in the future. Once again this is not likely to affect the decision to go ahead unless we are confident that an option will have a significant commercial value at the end of its life.

8.9.6   In-service development

When planning and analysing a prospective project and its development it is usual to assume that the development will be complete (and compliant with the specification) sufficient to meet the required service life. In practice equipment and plant often change during service operation.

The changes are usually required to meet new system requirements or to take advantage of new methods of working or of new technology. Although this will probably happen it is not known when or how. For these reasons it is usual to omit from the LCCA in-service development and system change or enhancement during its life: the LCC analysis is based on the system *as it is now envisaged* and future possible but unspecified change is ignored. Such changes in the future should be dealt with as a new development decision *at that time* using LCCA of the change as part of that decision process.

There are exceptions to this where, for instance, it is known at the outset that one of the system options employs technology that will become out-of-date (and inefficient or not economically supportable) during its life. In such cases an attempt must be made to forecast the re-development programme. DCF can then be used to compare it with other options that are able to survive the required service life, since the options will have different expenditure profiles.

8.10   LCC study implementation

The scope and content of an LCC study are wholly determined by the needs of the project and its project manager. The first and most important task is to establish the study 'philosophy' and to agree in advance the study objectives, approach, scope, leading assumptions and investment criteria. It is disastrous to try to address these matters at the end of the study as key decision dates approach and it is too late to take proper account of them.

This is followed by detailed discussions with specialist teams to ensure that adequate definitions of viable system options are available, including any do-nothing options. This leads to the creation of a model of the life cycle within which cost data elements can be aggregated and combined.

There is then a data gathering phase using data from analogous or prior projects and extrapolating to the new project or system situation. This can be a time-consuming task and may need specific data collection and analysis work when appropriate historical records do not exist.

Data gathering is bedevilled by the inadequacy of historic project records *for the purpose for which they are now required.* Apart from the technical differences of the new project, there are often problems of perception and definition. A cost can be a quite different thing to an investor, to a project engineer, to a maintenance manager, to an accountant or to a sub-contractor and data definitions must match or be made to match. It is also necessary to know whether reported costs are current or to which base year they refer, and what overheads are allocated (a difficult but very important factor) and whether profit is included or not. There is rarely such a thing as a simple straightforward 'cost' already available. It should also be recognised that some of the analyses are bound to be subjective and past data often missing or inadequately defined.

Once data collection is complete, the calculation of LCC profiles is usually a relatively quick task particularly when computerised models and tools are used.

The final part of the study is the most important of all: the communication of the results to the project manager and the project manager's responses to the option and sensitivity analysis. If the project manager is not properly informed then decisions could well be flawed. Note that it is not the role of the LCC analyst to make the decisions, but to advise.

It is unfortunate that the LCC analysis usually comes after other assessment activities and is dependent on other project members' work and their technical and feasibility studies. These studies and the data collection can often take up a large amount of the available assessment timescale. This leaves inadequate time for the essential dialogue with the decision maker and the re-run of alternative system and subsystem options. This is one of the reasons why computerised models can be so valuable. This final stage of the study must be very carefully planned at the outset and due time (with some contingency) allowed for other contributors' delays.

8.11 Value

So far, the cost of a project has been discussed. However, a project manager or decision maker is not necessarily seeking the lowest cost. The best value is sought, and cost is not necessarily the whole story. One of the system options may be particularly attractive even though it costs the same as, or more than, another option. It may be better aesthetically, match the culture of the organisation, be less disruptive, may be less sensitive politically or have a better planning or environmental acceptability.

Although it is possible to ascribe costs to these attributes, it is difficult and can be controversial (e.g. the inclusion of the 'cost' of eliminating a twelfth century church was used in one project analysis). It can also be misleading to include such items in the body of an LCC: it is better to deal with these matters outside the analysis as part of the broader decision process. However, the identification of such value factors can be a useful service to the decision maker even though they are not costed explicitly.

8.12   Uncertainty and risk

LCCA are usually attempting to forecast the future out turn of a project. Unfortunately the future is full of real uncertainty, particularly when development and technical innovation are necessary to the realisation of the project.

The uncertainty extends from technological uncertainty and relevance of the development and manufacturing organisation's past experience, to development and construction timescales. It also includes the reliability, maintainability and durability of the system in-service.

Thus, the calculation of a single 'most likely' LCC is only part of a much wider story. What the decision maker really needs is an indication of the possible variance in the LCC. The real uncertainties need to be explored in cost terms.

By judicious questioning of project development and management staff it is usually possible to set realistic upper and lower bounds on each of the elements of the LCC. The LCC model can then be used to explore the consequences of these ranges of parameters and the sensitivity of the results to them. More formal statistical risk analysis can also be performed using these ranges of uncertainty.

If the excursions of the model show that the ranking of system options is unaffected, then the cost decision is straightforward. If, however, the rankings change then the model shows which values of which parameters have a significant influence and identifies those areas where further analysis and refinement of the model would be useful. It also highlights the areas where management attention must be focussed and careful control will be necessary during construction, development or operation.

8.13   Resources

The LCCA involves the costing of all the resources that will need to be acquired and consumed during a project and brings everything to a single bottom-line figure. This figure is very important since it identifies the money that will have to be spent in undertaking, supporting and operating the project.

However, different options may consume quite different mixes of manpower and material resources. The project manager may be as concerned about the acquisition of the resources, some of which may be less readily available, as about their cost. Two options with the same bottom line cost may not be equally attractive. Another important resource that is consumed is time. Time can have a very important cost impact, particularly on the development or production phase of a project's life. Another important and expensive resource is project finance.

It is also important to remember that just because a resource element is included in the LCCA does not necessarily mean that it can be bought when required, in the quantities wanted, with the performance or skills wanted or at the rate wanted. It may not be able to be obtained at all. It is most important to ensure that the model uses realistic values of all its elements. There is nothing that destroys the credibility of an LCCA faster than an unrealistic assumption about one of its parameters.

8.14 Calculation of life cycle costs

The calculation of LCC has already been identified as a modelling activity that represents the profiles of expenditure for each phase of a system's life and a combined profile for the total LCC itself. It has also been emphasised that life cycle costs arise from the expenditure, consumption or use of resources. The model of the LCC is, therefore, a process for accounting for all these resources, profiling them and then converting them to cost.

There are two broad methods that can be adopted for this modelling and calculation process which can be summarised as top-down or bottom-up.

The bottom-up models use explicit engineering, programme and support elements and activities to create a high fidelity model of the life cycle, and the phasing and all the interrelationships of the elements with one another. These models tend to concentrate in most detail on the operation and support of the system. They are often created explicitly for the project and its particular characteristics and relationships, and are characterised by the need to input relatively large quantities of data on such things as operating cycles, reliability, repair times and spares provisioning.

There are also available computer-based 'framework' models. These are general purpose bottom-up models where much of the work is already done with aided data input, conversion and profiling, and pre-formatted output formats and reports. Pre-programmed algorithms may be provided for some of the inter-element relationships and ranges of uncertainty and risk. The model may be designed for direct interface with project management (activity network) tools and may also include internal routines for availability modelling, discounted cash flow, spares scaling and maintenance optimisation. These latter facilities are used for option studies and non-cost optimisations before final costings are produced.

These framework models also need 'throughput' data on such things as build cost or development and production costs, together with the costs of maintenance and support equipment, and repair facilities. The throughput costs will be obtained by commercial quotations or from price lists or will themselves be generated by traditional engineering build-up (bottom-up) or high-level parametric costing methods.

Top-down or parametric LCCA models take a different approach and are particularly useful in the earlier phases of projects. These models can be used to predict likely development and production costs for hardware and software projects as soon as their general broad characteristics are established. These characteristics also influence the costs of subsequent operation and support.

In contrast to the detailed engineering element approach, the top-down models work from a limited number of indirect parameters. Such parameters as performance characteristics, operating requirement, application, and design and technical complexities are used to characterise the project. The models then use embedded relationships (cost estimating relationships) derived from the analysis of earlier analogous projects and past experience to create probable resource and cost values for the new project.

A number of such models are available commercially, each with its own strengths and special facilities. Some have been developed from common databases or share some core algorithms. These commercial models are available for personal computer use (or occasionally by timeshare remote access).

8.15 Summary

A life cycle cost analysis or study is not primarily about costs but about resources: material items (hardware, software), personnel, finance and time. It concerns all the things needed to acquire, purchase, deploy and use to get the project into service and to then run it for the remainder of its service life.

A life cycle cost analysis has a major secondary benefit in that it is one of the parts of the project evaluation process where a whole system view is taken of the procurement of a system and its operation. It is thus a consistent vehicle for trading-off all parts of the system among all phases of its life.

A life cycle cost analysis is a major element in the project decision-making process that allows a project manager to determine the cost consequences of all the technical, schedule and procurement options.

9 The way ahead for life cycle costing in the construction industry

J.W. BULL

9.1 Introduction

In its most rudimentary form, the life cycle costing (LCC) or whole life cost of a product, including construction industry building structures, considers the four following elements:

1. Purchase cost.
2. Maintenance and running cost to ensure effective performance.
3. Cost of in-service failure.
4. Disposal cost.

For a simple product such as a pencil:

1. The purchase cost is small, less than £1.00.
2. The maintenance cost is a pencil sharpener, perhaps £1.00, which can be used for many hundreds of pencils.
3. The cost of product failure is nil, as another pencil is usually immediately available.
4. The disposal cost is again nil, i.e. the nearest wastepaper bin.

The useful life of a pencil is about three months.
   For a product such as an automobile:

1. The purchase price may be between £10 000 and £20 000.
2. The depreciation, maintenance and running cost can vary between £3 000 and £8 000 per year.
3. The cost of product failure is the hire of a replacement vehicle.
4. The disposal cost may be a credit if the automobile is sold on.

The product's life will be eight to ten years, with a reducing depreciation cost, but with an increasing maintenance cost and an increasing chance of automobile failure.
   For a product such as a passenger aircraft:

1. The purchase cost may be tens of millions of pounds.
2. The maintenance and running costs are high, running into millions of pounds per year due to high energy input and labour intensive safety and servicing requirement.

3. The cost of aircraft failure may be tens of thousands of pounds per day in lost revenue if the aircraft cannot be used, or has to be diverted while in flight, to tens of millions of pounds if the aircraft crashes with loss of life.
4. The disposal cost may be a credit if the aircraft can be sold on.

The life of a passenger aircraft will depend upon safety requirements and profitability, leading to a life of twenty years up to forty years.

In considering the three examples, it becomes clear that as the initial cost of the product increases, so does the combined total of depreciation, failure, running and maintenance costs. In the case of the automobile and the passenger aircraft, the cost of running and maintaining the product is used to determine the point at which the product should be replaced. Thus the reasons to use some form of LCC analysis increases as the initial product cost increases.

9.2   Further considerations

In many areas of business, LCC is the norm. For example, defence procurement, electronics, mechanical engineering components and transport operations are all subject to increasing LCC evaluation. In these areas, it is very clear who the product seller is and that the seller has a reputation to protect. The seller hopes the purchaser will place further orders with them. The purchaser may have a strong market position over future orders and be seen to be clearly responsible for the maintenance and running of the product. In short, product failure could cause financial problems for both the seller and the purchaser.

In the case of the building and civil engineering construction industry, the capital cost of construction is almost always separated from the cost of maintenance. The cost of disposal or demolition is rarely a design consideration. It is normal practice to accept the cheapest capital construction cost and then hand over the building structure to others to maintain. This approach to cost may be acceptable when the expected life of a building structure such as a house is eighty years or where the building structure will remain substantially unaltered during its lifetime as, for example, a warehouse. However, once people or businesses begin to use the building structure, a range of other costs become important. Businesses now expect to change and adapt a building every ten to twenty years, with a major refurbishment after thirty years.

There have also been changes in the way a building is instigated. A property developer may lease or purchase land, then instruct consulting engineers to design and construction companies to build the structure. An estate agent may be asked to let or lease part or all of the building structure. Businesses using the building structure may be able to gain excellent contract conditions which enable them to use the building structure for purposes for which it was not designed. The property developer may well sell on his interest in the building structure, the estate agent may change, the contractor may have ceased trading, etc., thus the legal areas of responsibility for maintenance, repair, safety and upgrading of the building structure may become a subject of litigation. All of this means that people or businesses using the building will, in the future, expect clearly defined contracts over who is responsible for maintenance, upgrading, etc. The businesses may require guarantees in this respect and LCC analysis will be used as part of the contract conditions. Demolition and disposal costs are often included in the cost of a replacement development or are deducted from the value of the land. This may be sensible for such building structures as houses or offices, but in the case of certain high safety requirement industries, the cost of demolition and disposal may be even greater than the construction of the replacement building structure and thus must be integrated with the initial LCC analysis.

In the case of structures owned by local government, the capital and maintenance budgets are kept separate, but a life expectancy based on existing and likely future use is estimated. Consider, for example, the case of a highway bridge when gross vehicle weights are legally increased. Maintenance and repair costs are increased and the life expectancy of the bridge is reduced. In this case, if in the LCC analysis, at the design phase, an increase in gross vehicle weights had been considered, it would have been accepted that to spend an additional 10 to 15% on the capital construction cost would result in a long-term reduced maintenance cost, with no additional upgrading cost. The life expectancy of the bridge would then not have been reduced.

What is needed in the construction industry is an LCC approach to the purchase cost, the maintenance cost, the running cost, the cost of in-service failure and the demolition and disposal cost of a building structure.

9.3 Summary of chapters

9.3.1 Some methods used in LCC analysis

In chapter 3, a specific example of the use of LCC analysis in the refurbishment of a building is described. Two methods are used to optimize, in monetary terms, for the lowest LCC. The first method calculates the LCC for a number of options thus identifying the lowest LCC strategy. The second method uses mixed integer programming and vector algebra to determine a mathematical model of the LCC. The second model can be limiting in its use as it requires some detailed understanding of certain areas of mathematics. With the greater accessability of personal computers, the mathematical approach will be increasingly adopted.

The safety or serviceability of a building structure is often evaluated using reliability theory. Reliability theory uses a considerable amount of linear mathematics, non-linear mathematics and dynamic programming techniques. Chapter 2 discusses reliability-based design as used to optimize the LCC analysis. This means that safety is expressed as a reliability or as a probability of failure and is related to the risk a society is prepared to accept should the building structure fail in service. Serviceability, on the other hand, is a function of structural behaviour and is related to loading and to section properties. Although serviceability has a meaning similar to safety, serviceability is more difficult to define. Relating LCC analysis to serviceability is at best difficult.

9.3.2   Practical examples of LCC analysis

Chapter 4 considers the use of LCC analysis in the appraisal of highway projects. Having accepted that a highway should be built, the short-term view would be to accept the lowest design standard solution which has the lowest capital cost, i.e. just building a highway. This solution ignores the reason for the highway, which is to provide a service. The use of LCC analysis will identify that to accept a higher initial design standard and a higher capital cost will result in a lower future cost in terms of maintenance and upgrading. Further, if the LCC analysis includes the strategic issues of national road planning and the cost to the road user of delay, vehicle maintenance and accidents, then even higher capital costs will be justified, as road transport costs are often a significant portion of the GDP for many countries. The LCC analysis of highway schemes is now accepted by the World Bank and the importance of LCC analysis will continue to increase as the interrelationships and the analysis models evolve.

Chapter 5 considers LCC analysis in the procurement and operation of defence equipment. The use of LCC analysis in the defence industry is necessary because of the extended life expectancy of defence equipment and the high support cost of ensuring that the equipment stays effective. The defence industry is strictly regulated and the data necessary for LCC analysis is readily available, constantly updated and used to forecast, with great confidence, future support costs. In the defence industry, initial costs are rapidly rising and the systems are expected to last longer. With extended in-service life, the cost of scheduled and especially unscheduled support becomes increasngly significant. As there is usually only one initial purchaser for defence equipment, each defence requirement can be carefully considered and the resource allocation in the form of required public money fed in right from the start. This allows the principal cost drives to be clearly identified at an early stage. A mid-life update or refurbishment can be built into the LCC analysis. Further, there are often models and prototypes built and tested. The cost of documentation, training aids, etc. are all included. The time from concept to delivery may take ten years, with total delivery being spread over another ten years. A rather interesting problem not usually encountered in the construction industry is that defence expenditure is allocated on a year by year basis. Any yearly underspend is lost, thus sudden spending to a year's limit, at the end of a financial year may well alter a LCC analysis! Another aspect is equipment reliability. If reliability is increased, the procurement requirement is reduced and the defence industry profits are reduced accordingly. Consequently, a relationship is quickly established between increasingly the equipment cost to ensure reliability and the savings generated by purchasing fewer items of equipment. LCC analysis is effective in the defence industry, as there is one major purchaser with a well developed long term requirement, a stable source of finance and a need to ensure a future defence industry. The construction industry has much to learn from the defence industry.

9.3.3 Difficulties to be considered in LCC analysis

Chapter 7 shows that the principles of LCC analysis and the associated evaluation methods can be readily demonstrated in theory, but that there are considerable difficulties in practice. These difficulties are often associated with a lack of knowledge on the part of the LCC analyst, a lack of understanding on the part of the client, the uncertainties of the historic data, the time horizons and the policy issues of asset management. The use of LCC analysis on past projects helps to identify which data sets need to be assembled and what the appropriate characteristics of those data sets should be. New technologies affect many aspects of society. The values in a society change and these changes have an effect on LCC analysis. One of the requirements of LCC analysis is the education of those responsible for construction projects. They should be encouraged to consider the future effects of their present requirements. The owners and users of building structures must be educated into what the building structure is designed to do or can be made to do and how to obtain the best from their building structure. A logbook of work done on a building structure, plus a maintenance handbook would provide a basis for comparing predicted, actual and achievable LCC optimums.

Chapter 1 considers that it is wise to forecast the consequences of decisions, because the construction industry is usually using other people's money. It is necessary to ensure that all the interested parties to a construction project are psychologically prepared to accept the LCC analysis and if the analysis is proved to be wrong, to say, 'everybody said we were right at the time!' If, however, after completing a LCC analysis, a subjective decision, such as 'change the design, it does not fit the company image' destroys the LCC analysis, 'value engineering' must then be used to reassess the inputs and objectives of the LCC analysis.

Essentially LCC analysis is an auditable financial ranking system for mutually exclusive alternatives which can be used to promote the desirable and eliminate the undesirable in a financial environment. Often the values of the multiple variables in the LCC analysis are uncertain and a range of values are used to obtain a sensitivity analysis. Sometimes the best that can be achieved is the establishment of probabilities related to a series of final solutions. Often LCC analysis can identify areas of construction practice which if tackled in a different way, would lead to additional savings.

The institutional investor may wish to purchase a building from a developer and will be looking for a high yield, high retained value, low management costs and happy and secure tenants. The developer will want to make money fast, will attempt to buy low, attract users quickly, sell high and increase the yield on the investment. Most business people choose to lease a building structure as a building is only a means to an end. The business wants the cheapest, functional solution now, especially if the business has high resource costs. A business person wants to use their funds for the mainstream business. The public sector uses LCC analysis to justify expenditure from the public purse and is looking for building functionality, auditable decision making and cost effectiveness.

Chapter 6 discusses difficulties inherent in implementing LCC analysis. Many of these difficulties are associated with resource allocation, deciding on priorities and the system within which the decision-making process has to function. Often, within the decision-making system, it is assumed that reduced material usage means reduced costs, but this ignores constructability and changes in future use. Further, the effects of the decision-making process on total costs reduces as construction progresses. Decisions, based on LCC analysis, made at an early stage in the project, are more effective than the same decisions taken at a later stage, say when construction has started.

Money spent on LCC analysis is quickly recuperated. In the case of academic and teaching hospitals, the equipment cost can exceed 70% of the construction cost. Construction must proceed on the basis that equipment finance will be available. If equipment finance is suddenly not available, then there is no reason for the construction of the hospital! This late decision making can completely nullify a LCC analysis. In the end, the LCC analysis of an academic and teaching hospital is related to the effect on national productivity of having a healthy population with access to adequate health care. The decision-making process is then related to the existing national financial and national human resources, and to the future costs and the future benefits to the nation. After all, the health-care quality of life is life cycle costing!

9.3.4 LCC analysis as a decision tool

Chapter 8 considers LCC analysis as a decision aid and the attempts to model the acquisition and operating processes in terms of resources consumed. The resources consumed are then converted to a single baseline of total cost, cost profile and future cash flow. Correct scheduling of the consumed resources can significantly reduce total cost and maximize the useful work done. The requirements of the client plus any legal requirements must be clearly understood and met. It is useful to cost the do-nothing option, as this gives a baseline from which to work. The level of accuracy of the LCC analysis will vary depending on the availability of data, the industry for which the construction is required and the present state of evolution of the LCC process. It is also assumed that the required resources will be available as planned. However, the LCC analysis should make provision for the delayed availability of resources.

A point often overlooked is the communication of the results of the LCC analysis to the final decision maker. The LCC analyst should ensure that the decision maker understands what is being presented, the possible variance within the options considered and that sensitive questions within the agreed philosophy are asked and answered.

9.4 Considerations and conclusions

A major part of the difficulty associated with the full introduction of LCC analysis into the construction industry is related to the fragmented nature of the industry itself. Each stage of the plan, design, build, use, maintain and demolish process is very often considered by the client as a series of separate financial entities. This assumes that there is only one client! The use of LCC analysis in the plan and design stages increases the cost of those stages. The cost increase is usually deplored by the client, even though the client may have a much reduced cost during the remaining stages of the construction process. Yet a client may consider it perfectly normal for LCC analysis to be applied to the equipment and plant used on the construction site.

Equipment and plant suppliers use LCC analysis to guide them into achieving certain minimum income requirements over the life of their equipment and plant, otherwise they will go out of business. The equipment and plant suppliers have an advantage in that the products they offer for hire or sale have known capabilities, service requirements and life expectancies, and are used for clearly defined tasks. This is, unfortunately, not the case for the main products of the construction industry, namely the building structures themselves. Much of the reason for this is that although there is often a clear idea of who the seller is, perhaps with a long-term reputation to protect, there is little opportunity for the purchaser to come back with more orders. The purchaser may only require one building structure every twenty years and thus will not be in a strong market position to influence future construction orders. It may not be clear to either the seller or the purchaser who is responsible for the maintenance and running of the building structure. This is in contrast to the defence industry, where there is one very major purchaser with a well developed and thought out long-term requirement. The defence industry has a stable source of government-provided finance and a need to ensure a viable defence industry in the future. The construction industry could learn much from the defence industry.

Another potential problem area is related to the need to ensure that all the parties to a construction project are psychologically prepared to accept the LCC analysis and its implications. The LCC analyst should ensure the decision maker understands what is being presented, the possible variance within the options considered and that sensitive questions within the agreed philosophy are asked and answered. A complex LCC analysis can be destroyed by the client making a subjective decision which should have been considered in the light of value engineering related to the inputs and objectives of the LCC analysis.

The principles of LCC, which in itself is an auditable financial ranking system, and the associated evaluation methods, can be readily demonstrated in theory. In practice, there are considerable difficulties in using the technique as the values of the variables may be uncertain and the best that can be achieved is the establishment of a series of probabilities. The uncertainty of the variables is related to the lack of knowlege on the part of the LCC analyst, the lack of understanding on the part of the client, both of which are linked to the uncertainties of the historic data, the often undefined time horizons and the present and future policy issues of asset management. For example, in the construction industry the safety or serviceability of a building structure may be evaluated using reliability based theory and design which is then used to optimize the LCC analysis. This means that the safety or serviceability of a building structure is related to the risk that a society is prepared to accept should structure fail in service. What society is prepared to consider as an acceptable risk today may change in the future and hence be linked to the lack of knowledge on the behalf of the LCC analyst.

Many of the difficulties in LCC analysis are associated with resource allocation, deciding upon priorities and the system within which the decision-making process has to function. For example, the military defence area is strictly regulated and the necessary data required for a LCC analysis are readily available and constantly updated. It is easy for a relationship to be established between increasing the cost of defence equipment to ensure reliability and the savings generated by having to purchase fewer items of equipment. In the construction industry, this precision of data just is not available.

What can be done in the construction industry is to use LCC analysis to model the acquisition and operating processes in terms of the resources likely to be consumed. These resources can be converted to a single baseline of total cost, cost profile and future cash flow. As the correct scheduling of the resources can significantly alter the baseline of total cost, the LCC analysis would ensure that the useful work done is maximized. It must be remembered that correct decisions, based on LCC analysis and taken at an early stage of a project, are much more effective than the same decisions taken later in the project schedule. Correct decisions save money and increase profits.

With the increased accessability of personal computers and the increasing sophistication of LCC analysis software, there is now every opportunity for each building structure to be analysed in terms of LCC. The LCC analysis allows a number of options to be considered and the lowest LCC strategy to be selected.

Index